中国民居建筑艺术

The Architectural Art of the Traditional Vernacular Houses of China

陆元鼎　陆　琦　著
by Lu Yuanding and Lu Qi

中国建筑工业出版社
CHINA ARCHITECTURE & BUILDING PRESS

寿

禄

图书在版编目(CIP)数据

中国民居建筑艺术/陆元鼎，陆琦著．—北京：中国建筑工业出版社，2011.2
ISBN 978-7-112-12794-8

Ⅰ．①中… Ⅱ．①陆… ②陆… Ⅲ．①民居-建筑艺术-研究-中国-汉、英 Ⅳ．①TU241.5

中国版本图书馆CIP数据核字（2010）第260671号

翻　　译：黄运亭　杜可君
摄　　影：陆　琦
策　　划：张惠珍
责任编辑：李东禧　唐　旭　马　彦
责任校对：关　健

中国民居建筑艺术
The Architectural Art of the Traditional Vernacular Houses of China
陆元鼎　陆　琦　著
*
中国建筑工业出版社出版、发行（北京西郊百万庄）
各地新华书店、建筑书店经销
北京方舟正佳图文设计有限公司制版
北京画中画印刷有限公司印刷
*
开本：880×1230毫米　1/16　印张：30　字数：928千字
2010年12月第一版　2010年12月第一次印刷
定价：298.00元
ISBN 978-7-112-12794-8
(20105)

前言
Preface

中国历史悠久、文化灿烂，是一个伟大的多民族国家。中国古建筑，如宏伟的万里长城，横贯南北的大运河，北京故宫，各地著名的坛庙、寺观、园林等，早已闻名于世。可是，遍布各地的、与老百姓生产生活密切相关的、多姿多态的优秀民间宅居、村落、集镇却鲜为人知。

民居建筑是人类最早、最大量，与人类生活最密切相关的建筑类型，也是人类最原始、最持续发展的建筑类型。民居在一定程度上揭示出不同民族在不同时代和不同环境中生存、发展的规律，也反映了当时、当地的经济、文化、生产、生活、伦理、习俗、宗教信仰以及哲学、美学等观念和现实状况。各地区、各民族人民在民居建造过程中，都根据自己的生产生活需要、经济能力、民族爱好、审美观念，因地制宜、因材致用地进行设计和营造，有着极其丰富的经验。因而我国的传统民居建筑既有重要的历史、文化、科学价值，又有艺术欣赏和技术应用价值，它是我国民间建筑中一笔极其宝贵的财富。

为了面向爱好中国传统建筑文化的世界各国人士和专家，特编辑本书介绍中国民居建筑艺术。

China is a great multi-ethnic country with a long history and splendid culture. The ancient architectures in China have long been known to the world, among which are the magnificent Great Wall, the Grand Canal traversing north-south, the Forbidden City in Beijing, and famous shrines, temples, gardens around the country. However, little known are brilliant vernacular houses, villages, market towns with great varieties, which are found all over the country and closely related to the daily life and production of the ordinary people.

As the earliest architectural type in man's history which boasts its largest quantity and is inseparable from man's life, vernacular architecture is also the most primitive one with sustainable development. To some extent, it not only reveals the surviving and developmental rules of different ethnic groups in different eras and environments, but also reflects the concepts and realities of that period in terms of local economy, culture, production, life, ethics, customs, religion, philosophy, aesthetics and so on. Based on their production and living need, financial abilities, preferences, aesthetic concepts, in construction, different ethnic groups in various regions have experience in design and creation of their own buildings in line with the local conditions and materials and have accumulated rich experiences. And therefore the traditional vernacular architecture in China is of important historic, cultural and scientific value as well as the value of art appreciation and technology application. It is a precious treasure of the Chinese folk architecture.

This book is an introduction to the Chinese vernacular architectural culture, devoted in particular to foreign friends and experts who are fond of the traditional Chinese architectural culture.

目录

C O N T E N T S

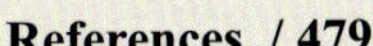

一、中国民居建筑的历史与文化

Ⅰ. The History and Culture of the Chinese Vernacular Architecture

（一）中国民居建筑的产生与发展

衣食住行是人类生存的四大基本要求，为了满足住的需要，于是产生了居住建筑，它是历史上最早出现的建筑类型。

远古时期，原始人居住在天然洞穴之中，称为“穴居”。在南方，因气候湿热，野外多虫兽，于是在树上构筑简陋的窝棚栖息，称为“橧巢”，也称“巢居”。其后发展到半地下，再到地面，这是早期的居住建筑，在我国陕西半坡有大量的遗址为证。

在长江下游，浙江余姚河姆渡发掘出的聚落遗址，是一座长度在30米以上的干阑式长屋，进深约7米，前檐宽约1.3米，建于木桩上。长屋地板比地面高约1米，用木梯上下。其梁柱等构件交接处的榫卯构造已相当复杂，木构建筑技术在当时是一个大进步。

封建社会开始，以农业生产为主。根据墓中出土的画像石、画像砖和明器、陶屋，可以推知汉代小型住宅的平面为方形、长方形，稍大的住房有曲尺形，内部有院落，正中有楼高起，次屋则低矮，外观主次分明。贵族的大型宅第有几进院落，正屋旁还有辅助房屋、宾客住房。院落内前堂为主要建筑，后堂有屋，是古代前堂后室制的发展。

在北魏、东魏时期留下的石刻中看到，贵族住宅有大型厅堂和庭院回廊，甚至在住宅后部建有园林，是园林式住宅的初创阶段。

隋唐五代的民居建筑，仅能从敦煌壁画和其他绘画中得到一些旁证，如贵族宅邸大门采用乌头门形式，宅内两座主要房屋之间用带有直棂窗的回廊，连接成四合院。

宋代农村住宅可见于宋画《清明上河图》中，茅屋比较简陋，墙身较矮，有些是茅屋和瓦屋

i. The origin and development of the Chinese vernacular architecture

Clothes, food, accommodation and transportation are the four basic necessities of man's life. To meet the needs of accommodation, housing architecture came into being and became the earliest architecture type in the history.

In the ancient times, the primitive people dwelled in natural caves, which were referred to as cave-living. In the South, due to the hot and humid climate, insects and wild animals in the open country, the crude shanty in the trees became the habitat, known as the nesting. Later, people began to build their habitat half underground, and then on the ground. A large number of these earliest residential sites have been found in Banpo of Shaanxi Province in China.

In the lower reaches of the Yangtze River, ruins of a village were excavated in Hemu Ferry, Yuyao, Zhejiang Province. It is a long bole-fence house stretching more than 30 meters, built on wood piles with the spatial depth of around 7 meters and front brim of 1.3 meters wide. The house floor is about 1 meter higher than the ground, with a wooden ladder for access. The tenon and mortise joints connecting other components such as beams have a complex structure, indicating that there was a big step forward in the wood construction technology at that time.

The feudal society gave priority to agricultural production at its early stage. From the unearthed painting stones, bricks, afterlife utensils and pottery houses from the tombs, we can infer that the small houses in the Han Dynasty were usually of square or rectangle shape. Relatively larger houses were designed in tri-square shape, having a courtyard inside, with the high building in the middle and sub-houses being relatively low so that the contour of the residence has a clear distinction between the major and the minor houses. A large-scale aristocratic residence is composed of several courtyards and besides the main building, there are rooms for miscellaneous purposes and guests. Inside the courtyard, the front hall is the main building and rooms are placed in the rear hall, which is developed from the ancient front-hall-rear-room principle.

The stone carvings in the Northern and Eastern Wei periods show that the aristocratic houses are built with large halls and corridors, with even gardens in the rear of the residence. It is the original form of the garden house.

The vernacular architectures of the Five Dynasties and the Sui and Tang dynasties can only be found in the Dunhuang mural paintings and some other paintings. For instance, the entrance door of an aristocratic mansion adopts tortoise-head shape; two major buildings inside the

相结合构成的一组房屋。城镇小型住宅多用长方形平面，屋顶为悬山式或歇山式，除草葺和瓦葺外，山面的两厦和正面的披檐则多用竹篷或屋顶上加建天窗。稍大的住宅，外建门屋，内部采用四合院形式，有些在院内栽花植树，美化环境。

此外，宋代王希孟的《千里江山图》中所绘住宅多所，都有大门、东西厢房，主要部分是前厅、穿廊、后寝所组成的工字屋。这种在古代前堂后寝制的传统布局原则下，中间用穿廊连成工字形、王字形平面的布局，成为了本时期民宅布局的一大特点。

明清时期，人口骤增，经济兴旺，文化发展，城乡面貌更显繁荣，本时期有大量的民居建筑实物至今尚存，并遍布城乡各地，它基本上仍保留着各地各民族民居的传统特色。

(二) 中国民居建筑与文化

中国历史悠久，土地辽阔，文化遗产极为丰富。建筑文化遗产中的最大量的、与广大人民生产生活密切相关的传统民居，也同样非常丰富。今天，它仍然散布在全国各民族和各地区，虽经风雨沧桑，仍然为广大人民所使用。其中不乏优良的传统精华，艺术精湛，是我国珍贵的文化宝藏。

我国南北气候悬殊，山丘河海地理条件各不相同，材料资源也有很大差别，加上各民族不同的风俗习惯、生活方式和审美要求，导致我国民居建筑具有鲜明的民族特色和丰富的地方特性。

中国传统民居建筑与社会、历史、文化、民族、民俗有关，又与儒礼、道学、阴阳五行等思想学说有密切联系。优秀的传统民居建筑具有历史文化价值，同时，又有实用和艺术价值，这是我国宝

residence are connected with cloister with straight frame windows, forming a courtyard.

Rural dwellings in the Song Dynasty can be found in the famous painting of that time, Along the River During the Qingming Festival, thatching huts being relatively simple and primitive, with low walls, and a group of cottages and tile houses. Small houses in cities and towns are of rectangular shape, having overhanging gable roof or xieshan-style roof (roof with four sloping surfaces and nine hips). In addition to grass and tile coverage, two sides of the roof and front eaves use bamboo coverage or adopts skylight roof. For a larger house, a door-house is built outside and inside is built in the form of courtyard with trees and flowers planted for landscaping.

Besides, in his famous painting of A Thousand Li of Rivers and Mountains, Wang Ximeng of the Song Dynasty painted some houses and each has an entrance door and chambers in eastern and western wings. Yet, the main part of the residence is a 工-shape house consisting of the front hall, corridor and rooms at rear. The 工-shape or 王-shape plane connected by a corridor in the middle, which was built in accordance with the traditional layout of a front-hall-rear-room in ancient time, becomes a layout feature of the vernacular house in that period.

During the period of Ming and Qing Dynasties, a surge in population, economic prosperity and cultural development brought about the prosperity in urban and rural areas. Considerable vernacular architectures all over the nation, retaining the traditional characteristics of various ethnic groups in different regions, have survived to this day.

ii. The Chinese traditional vernacular architecture and culture

China has a long history, vast land and rich cultural heritage. As the most substantial architectural culture heritage are closely related to the production and daily life of the ordinary people, the traditional vernacular house is equally abundant. Today, it is spreading in different ethnic groups and regions all over the nation. After the vicissitudes of the time, it is still widely used by the ordinary people and some have shown us the quintessence of the fine tradition and exquisite art. It is a precious cultural treasure of China.

Apart from the disparities in climate in north and south China, different geographical conditions, material resources, customs, lifestyles and aesthetic requirements of various ethnic groups have resulted in the distinctive ethnic features and rich local characteristics in the Chinese traditional vernacular architecture.

贵的文化遗产，也是世界文化的珍贵财富，亟需保护和弘扬。

哲学是文化思想的集中体现，是民族精神的精华，是人类智慧的最高创造。在中国传统文化系统中，哲学处于核心地位，起着主导的作用。

从哲学看，中国传统文化主要受下列思想影响，即儒学、道学、阴阳五行学以及其后的中国佛学。他们相互影响、吸收、融合，构成了中国传统文化的总体。

中国传统文化的基本精神涵盖四个主要方面：以人为本的人文主义价值系统，自强不息、豁达乐观的民族心理，观物取象、整体直觉的思维方式以及天人合一的审美理想。就价值系统而言，中国文化表现了突出的以人为本的实用理性精神。

中国传统文化的思维方式有两个特点，直觉体悟的直观性和观物取象的象征性。天人合一是中国传统文化的审美理想和最高境界。

中国传统民居是中国传统文化在建筑中的体现。通过中国传统民居的布局、形制，我们可以形象而深刻地感受到中国传统文化思想的深广影响。中国传统民居充分体现了中国传统文化中的哲理观、宗法观、环境观、思维观，从不同层面反映了中国传统文化的渊博精深和高明智慧。

中国传统文化在中国民居建筑中的影响，首先是宗法观，即宗法礼制。宗法观的核心是等级观念，它对民宅中的布局、形制，甚至开间、规模、装饰、装修都作了严格的规定，目的是维护以血缘为纽带的封建等级制度。

其次是天人合一观念，它是民居择向、定位、选址的依据，又是汉族儒家、道家思想在大环境条件下处理建筑的准则，它包括建筑平面布

The Chinese traditional vernacular architecture ties up with the society, history, culture, nationality and folk-customs, as well as the theories of the Confucian rites, Yin-Yang and Five Elements of Taoism. The outstanding traditional vernacular architecture is of historic and cultural value and of practical and artistic value as well. It is not only the precious cultural heritage of China, but also the precious cultural wealth of the world, calling for urgent protection and promotion.

Philosophy is a concentrated expression of culture, the quintessence of the national spirit and the acme of human wisdom. In the system of the traditional Chinese culture, philosophy is at the core position, playing a leading role.

From a philosophical perspective, the traditional Chinese culture is influenced by the following ideologies, namely, Confucianism, Taoism, the theory of Yin-Yang and Five Elements and Chinese Buddhism in later time, which have absorbed, impacted and integrated with each other, forming the collectivity of the traditional Chinese culture.

The basic spirit of the traditional Chinese culture covers four main aspects, that is, people-oriented value system of the humanism; self-improvement, open-minded optimism of the national psychology; perceiving the concept of the object from an overall intuitive way of thinking; as well as the aesthetic ideal of harmony between man and nature. In respect of the value system, the Chinese culture represents the spirit of practical rationality emphasizing the people-oriented aspect.

The traditional way of thinking has two characteristics, the direct view of intuitive understanding and the symbolic view of perceiving the concept of the object. Harmony between man and nature is the aesthetic ideal and the highest realm of the traditional Chinese culture.

The traditional vernacular architecture of China is the embodiment of the traditional Chinese culture in the architecture. Through its layout and shape, we can feel the profound impact of the traditional Chinese culture and thinking. Mirroring the philosophy, patriarchal and environment concepts and thinking of the traditional Chinese culture, the Chinese vernacular architecture fully reflects, from different levels, the brilliant and profound wisdom of the traditional Chinese culture.

The impact of the Chinese traditional culture in the Chinese vernacular architecture, first of all, is the patriarchal view, that is, the family rituals. The core of the family rituals is hierarchy which has a strict requirement in vernacular houses in terms of their layouts, shapes, room spaces, scales, furnishing and decoration, so as to maintain the feudal hierarchy emphasizing blood ties.

The concept of harmony between man and nature comes in the second place, which serves as the basis for the orientation, location, and

局和空间组织结构的整体性、秩序性和教化性。在古代，中国传统民居天人合一的环境理想大多通过风水论说来实现，这就是五位四灵的环境模式，五位即东南西北中五个方位，四灵即四方神灵——青龙、白虎、朱雀、玄武。

再次是中国传统民居中的思维观，它充分反映了中国传统文化中的人本主义精神，一切以人为本，以民居建筑中的主人及其家庭为本，天地人三者结合，考虑的是人，把人的生产、生活作为民居营建的出发点。

这就是中国传统文化在中国传统民居中的影响和特征。

site selection of an architecture, as well as the principle for the building construction in a grand environment promoted by Confucianism and Taoism of the Han people. It includes architectural layout, integrity of space structure, order and domestication. In the ancient times, the ideal of harmony between man and nature in the Chinese vernacular houses was achieved through feng-shui, or geomancy, that is, the five-orientation and four-god environment mode. The five orientations are east, south, west, north and center; and the four gods in four directions are Black Dragon, White Tiger, Phoenix, and Xuanwu.

Then is the way of thinking embodied in the Chinese vernacular houses, which fully reflects the spirit of humanism in the traditional Chinese culture. Everything is people-oriented, prioritizing the owner and family of the house. The trinity of Heaven, Earth and Man has Man at the first place. Thus, the residential construction is based on man's production and living.

Those are the influence and features of the traditional Chinese culture in the Chinese vernacular houses.

二、中国民居建筑的构成与特征

Ⅱ. The Composition and Characteristics of the Chinese Vernacular Architecture

（一）中国民居建筑的构成

中国民居建筑的构成有四个因素，即社会因素、经济因素、自然因素和人文因素。

1．社会因素，包括生产力、社会意识、民族差异、宗教信仰和风俗习惯等。

我国是一个多民族国家。以汉族为例，它在历史上是一个长期以宗法制度为主的封建社会，家庭经济则以自给自足的小农生产为基础，并以血缘纽带作为宗族的维系，而维持社会稳定的精神支柱则是儒家伦理道德学说。这种学说提倡长幼有序、兄弟和睦、男尊女卑、内外有别等道德观念，并崇尚几代同堂的大家庭共同生活，以此作为家庭兴旺的标志。对民居建筑来说，它对内要满足生活和家庭生产的需要，对外则采取防止干扰的做法，实行自我封闭，尤其将妇女的活动严格限制在深宅内院之中。宗法制度的另一个重要内容则是崇祖祀神，提倡对宗族祖先的崇拜和对各种地方神祇的祭祀。这种宗法制度和道德观念对民居的平面布局、房间构成和规模大小有着深刻的影响。

中国封建制度的核心是等级制度和儒礼宗族制，如汉族院落式的民居平面布局就是这种形制的产物。封建制度等级森严，对于建筑的规模、大小、开间、进深以及屋顶形式，甚至装饰、装修、色彩都有严格规定。例如民宅不得超过三间，色彩规定为黑白素色，而大型宅第就可以多进、多院落，甚至多路建筑布局，并且可以带书斋、带园林。从民居的平面布局上就可以看到社会制度对建筑的影响。

2. 经济因素，这是民居形成的物质基础。

民居的营建需要材料，并要以一定的构造

i. The composition of the Chinese vernacular architecture

The composition of the Chinese vernacular architecture involves four factors, that is, social, economic, natural and cultural factors.

1. Social factors, including productivity, social consciousness, ethnic differences, religious beliefs, customs and so on

China is a multi-ethnic country. Take the Han people as an example. In the feudal society which gave priority to patriarchal system and dominated the history for a long time, the domestic economy was based on small self-sufficient production and the clan was maintained through blood ties. The spiritual pillar of the social stability was the ethics doctrine of Confucianism, which promoted the moral values of seniority, brotherhood, man being superior to woman, and differentiated tasks for men and women, and advocated the family pattern of several generations living together under one roof as a symbol of prosperity for the family. As for the vernacular architecture, it must meet the needs of living and domestic production and prevent the family from external interference at the same time. Such a practice of self-isolation imposed, in particular, severe restrictions on women's activities which were confined to the inner yard of the residence. Another important element of the patriarchal system is ancestor and god worship, worshiping the clan ancestry and offering sacrifice to the gods in the region. Such patriarchal system and moral values have far-reaching impact on the vernacular houses in terms of the layout, room structure and the scale.

Hierarchy and family rituals of Confucianism, the core of the Chinese feudal system, gave birth to the courtyard-style vernacular houses of the Han people. The hierarchy of the feudal system was so stern that it had hard and fast rules on the architecture in terms of the scale, size, width, spatial room, as well as the roof form, and even the furnishing, decoration and color. For example, the houses for ordinary people should have no more than three rooms, with only plain black or white color; while the layout for large houses could be designed with several blocks, courtyards and roads, and even with study and garden. The layout of the vernacular houses exemplifies the influence of the social system on the architecture.

2. The economic factor is the material foundation in the formation of the vernacular houses

Materials are indispensable for the construction of a vernacular house to be built in a certain configuration mode. Therefore, the quantity and quality of materials and their configuration determine the size, quality and grade of the architecture. The rich can decorate the doors, roofs and rooms in a very lavish way while the poor can only use mud walls and

方式建造起来，宅居所用材料的多少、贵贱和结构方式决定着民居建筑的规模、质量和等级。富有者可在建筑的大门、屋顶和室内进行华丽的装修，而贫穷者只能以泥墙挡风雨，薄瓦以蔽身。

然而，劳动人民的智慧是无限的。他们利用当地的材料，如木、竹、灰、石、黄土，根据当地的自然条件和自己的经济水平，因地制宜，因材制用，自由发挥，按照自己的居住需要和营造的规律来进行建造。因此，在他们的民居中，充分反映出了功能是实际的、合理的，设计是灵活的，材料构造是经济的，外貌形式是朴实的，这是建筑中最本质的东西（图1）。广大民居建造者和使用者是同一的，是自己设计、自己营建、自己使用三位一体的，因而，民居的实践更富有人民性、经济性和现实性，也最能反映本民族的特征和本地区的地方特色。

3. 自然因素，它包括气候、地形、地貌、材料等自然物质和环境因素。

我国东南部沿海，西北部为高山峻岭，南北总长约5500公里，东西长约5200公里。境内地貌西北高、东南低，西部有广阔的高原，东部有大片平原，其间分布着高山、丘陵，东南则多水网密布的河湖和溪流。

我国南北气候差异很大，北部寒冷，并有霜雪，南方地区夏季炎热、潮湿、多雨，有的地区终年不下雪。每年夏、秋季，东南沿海地区常有台风侵袭，台风来时还带有暴雨，对人畜、房屋伤害极大。

此外，由于地理、气候的不同，各地建筑材料的自然资源也有很大差别。中原及西北地区多黄土（图2），丘陵山区多产木材和石材，南方还盛

图 1 生土建筑 Adobe-earth architecture

thin tiles to build their shelters.

However, the wisdom of the working people is unlimited. With local materials such as wood, bamboo, ash, stone, loess, they adopt the construction rules to build houses to meet their own needs based on the local natural conditions and their economic levels. Hence, the vernacular houses represent the most essential features of the architecture, having practical and reasonable functions, flexible designs, economic materials and simple appearances. In particular, for most houses, the builders are also the users who design, construct and use these buildings. Therefore, the vernacular houses are people-oriented, cost-effective and practical, and most of all, reflect the ethnic features and the local characteristics of the region.

3. Natural factors include the climate, landform, terrain, materials and other natural substances and environmental factors

China borders vast seas on southeast and lies against steep mountains on northwest, extending about 5500 kilometers from north to south and covering a width of about 5200 kilometers from east to west. Within the territory, the high land in northwest lowers gradually southeastwards. In the west, there lies grandeur plateau; while in the east there is large plain. Mountains, hills, network of rivers and streams in the southeast are distributed in this wonderful land.

The climate in north and south China is quite different: In the north, it is cold with frost and snow; in the south, it is very hot, humid and rainy in the summer and in some areas there is no snow all the year round. Each year in summer and autumn, the coastal areas in southeast are often harassed by typhoons which bring heavy rain and cause great damages to people, livestock and houses.

Moreover, due to the diversified geography and climate, the natural resources for construction materials are quite different. Loess is mostly found in the central plain and northwest region; hilly and mountainous areas boasts prolific wood and stone; the southern region is rich in bamboo and brick; tile and natural gravel can be produced and collected in many places. In some coastal area, ash burned from shells can be found.

产竹材，许多地区都能自产砖瓦和采集天然的沙石，沿海还能生产由贝壳烧成的蜃灰。

民居建造是在一定的地点环境和一定的地理、气候条件下完成的。北方的气候干燥、寒冷，南方的气候闷热、潮湿，导致南北民居建筑的处理方式、手法都不一样。以地理环境来说，有坡地、平地、河流、小溪，民居建在坡地或平地上，或建于水畔，其景观效果都不一样。气候因素对民居建筑的平面布局、建筑造型以及内部空间影响更大，这也是不同地区民居形成不同模式、不同特色的重要原因。

4．人文因素，它包含民情、民俗、生产、生活方式以及文化、审美观念等内容。

汉族民居中，文化因素占重要地位，儒家伦理学说是主导思想，其中对民居建筑影响最大是崇天敬祖思想。

在民居设计中，祠堂、祖堂是建造房屋时首先考虑的内容。古制规定："君子将营宫室，宗庙为先，厩库为次，居室为后。"宋代朱熹所著《家礼》一书就有"立祠堂之制"，规定"君子将营宫室，先立祠堂于正寝之东，祠堂制三间或一间"，说明古制对祠堂之重视和限制（图3）。

祀宅合一的民居，建造时也是以祀为主。在设计时，先将供祀祖先、天地的场所作为祖堂，位置在整个宅第的最后一进的正中厅堂，称为后堂，亦称祖堂。后堂的开间、进深和脊高、檐高都有一定的尺寸规定，甚至神龛、香炉的位置、高度也有所规定，不得随意更改。

影响民居建筑的还有一种思想，即风水观念。

风水观，古称堪舆学，它来源于阴阳五行学

图 2 窑洞原土材料
Cave-dwelling original soil material

A vernacular house is constructed and completed in a given location and environment and under given geographical and climatic conditions. The dry and cold weather in the north and the hot and humid weather in the south lead to different approaches and methods in constructing vernacular architectures in different places. Taking geographical environment as an example, there are slope, flat land, river, creek, mountain and water. The houses built on the slope, flat land or beside the water present different views. The climatic factor has even greater impact on vernacular architecture in terms of the layout, design and interior space. That's why vernacular architecture in different places presents different forms and characteristics in various regions.

4. Cultural factors include folk situations, folk customs, production, ways of living, culture and aesthetic concepts

Cultural factors play a vital role in vernacular houses of the Han people, which is dominated by the Confucian ethics theory. The greatest impact on vernacular architecture is the thinking of worshiping the Heaven and ancestry.

The first consideration for the design of a vernacular house is the ancestral temple or hall. It is stipulated in ancient rite that for the house construction, priority should be given to the ancestral temple, second to the stall, then to rooms. In the chapter of Ancestral Hall Construction in Family Ritual composed by Zhu Xi in the Song Dynasty, it says for the house construction, the ancestral temple, in three-room or one, should be situated at the east of the main bedroom. It shows that the ancient rite attaches great importance to and restrictions on ancestral temples.

The construction of a vernacular architecture which is used for both living and worshipping gives priority to worshipping. In the design, the ancestral temple where the family's ancestry and the Heaven and

图 3 浙江东阳卢宅祠堂
The Ancestral Temple of the Lu's Former Residence in Dongyang City, Zhejiang Province

说，原是古代阳宅和阴宅在择位定向中考虑气候、地理环境的一门学说。阳宅即民居建筑，例如，在农村，对民宅的选址一般已形成一种比较固定的负阴抱阳模式，即村前要有流水，村后要有高山，房屋坐北朝南，地形前低后高。从现代观念来分析，这种布局原则还是有科学性的一面的。譬如村落面靠流水，这是食水、交通、洗濯的需要。村后高山作屏，可抵御寒风侵袭。地形前低后高，说明坡地上盖房子既要求干燥又要易排水，对居住及人体健康有益。

风水观念中还有一种象征和压邪思想，如江南、皖南一带民宅喜用马头墙。所谓马头墙，就是将山墙墙头部位做成台阶式盖顶，在盖顶之前沿部位，使墙头上翘成似马头形状，称为马头墙（图4）。据当地老人讲，山墙做成马头形状，说明该户家族中曾有人中举。武官用马头状，称马头墙，文官则用印章，方形，称印石墙。实际上，山墙做成马头墙或印石墙，是显示住户家庭中举、有朝官的一种用建筑表现的炫耀方式。老百姓家只能用双坡屋面。

广东潮州民居的山墙墙头部分有做成金、

Earth are worshipped is placed in the middle of the last block of the entire residence. It is named as the rear-hall, also known as the ancestral hall. There are restrictions for the width and depth of the rear hall and the height of ridges and brims, so are the position and height of an altar and incense burner, which should not be changed at random.

Another thinking that influences the vernacular architecture is feng-shui.

Feng-shui, or geomancy, originating from the theory of Yin-Yang and Five Elements, is a theory in the ancient China, comprising climate, geography and environment for the location and orientation of Yang-House and Yin-House, houses for the living or the spirit. The Yang-House is the vernacular house. For instance, in rural areas, the site of houses in general has a relatively fixed pattern, holding Yang and lying against Yin. That is, a village should face the running water in front and have a high mountain behind. A house should face the south; the terrain should fall forwards. Such a layout is reasonable by the analysis of the modern concepts. For example, the running water lies in front of the village, meeting people's needs for fresh water, transport and washing; the high mountain behind it can be a perfect screen to resist cold wind; building a house on a tilting terrain keeps the house dry and makes it easy for drainage, which is ideal for living and health.

Feng-shui contains symbolic and evil-suppression thinking. For example, the horse-head gable is quite popular for the vernacular houses in the region south of the Yangtze River and southern part of Anhui Province. The name given as the top of the gable is built into step-style canopy, which is shaped into a horse-head in the forefront roof, thus known as the horse-head stepped gable. According to the local people, where the gable is shaped into horse-head, it indicates that there has been a *Juren* (a successful candidate in the imperial examinations at the provincial level in the Ming and Qing Dynasties) in the family. For a military officer, the gable is shaped into the horse-head and named as horse-head stepped gable; for a civilian official, the

图 4 安徽民居马头墙
The piled horse-head walls of the vernacular houses in Anhui Province

水、木、火、土五行方式的，也是同样的道理。在实际调查中，民居建筑通常用两种山墙：一是曲线形，称水墙，另一种是金字形，称金墙。依照五行相生相克学说，水压火是五行相克论说，金生水、水克火，是五行相生又相克的论说，其目的和意图都是为了压火、防火。古代建筑因是木结构营造，最怕火灾，建筑一旦失火，无法可救，但当时科学水平有限，无法采取有效的防火措施，于是采用压邪这种祈望吉祥平安的心理手法，从而可见天地观念对民居建筑的深刻影响。

（二）中国民居建筑的特征

民居的特征，主要是指民居在历史实践中反映出来的本民族、本地区最具有本质的和代表性的东西，特别是要反映出与各族人民的生活生产方式、习俗、信仰、审美观念密切相关的特征。

具体来说，民居特征在建筑上主要反映在下列三个方面：

1．布局特征，主要表现为平面形式丰富和空间组合多变。

以汉族民居为例，大型者如多进院落式集居住宅（图5、图6），小型者如三合院或四合院住宅（图7），它们的基本布局都一样，前堂后寝、中轴对称、正厅两房、主次分明、院落相套，规整严谨，外部有高高的封闭围墙，内部则是层层院落，或纵向发展或横向发展，形成一种外封闭、内开敞，组合灵活而又紧凑的布局形式。在其他民族中，如云南傣族，在封建制度下，家庭盛行一夫一妻制，每户要收户头税，年长子女成家后要另立门户，故民居都属独户式竹楼。竹楼内是大空间，内

gable is built into seal or square shape and named as seal gable. All these shapes are actually show-off ways in the architecture to splurge that the family there has bred a *Juren* or an imperial official. For ordinary people, they can only use double-sloping roofs.

Likewise, in Chaozhou, Guangdong Province, the gable top is shaped in accordance with the practice of Five Elements of gold, water, wood, fire and earth. In the actual survey, two kinds of gables are usually adopted in the vernacular architecture: one is of curve shape called water gable; another is of pyramid shape called gold gable. In accordance with the Five-Elements theory, water suppresses fire while gold gives birth to water. Thus, the purpose of water and gold gable is to check and prevent fire. Since architectures in the ancient times were made of wood which was vulnerable to fire, once a building caught fire, it was impossible to put the fire out. However, because of the limited scientific knowledge, no effective measures could be taken for fire fighting, so people resorted to this psychological relief by checking the evil and praying for prosperity and safety. It well proves that the concept of Heaven and Earth has a profound impact on the vernacular architecture.

ii. The characteristics of the Chinese vernacular architecture

The characteristics of the traditional Chinese vernacular architecture mainly refer to the features most representing the people and the region, especially reflecting the production and ways of life, customs, beliefs, aesthetic concepts of various ethnic groups.

Specifically, the characteristics of the vernacular architecture are reflected in the following three aspects:

1. Layout, mainly manifested in the rich planar and varied spatial combination

For example, the vernacular houses of the Han people, whether a large house with multi-courtyard cluster or a small house with three-in-one or four-in-one courtyard, the basic layout is the same, featuring fore-hall-rear-room, symmetrical central axis, two-room and main-hall and marked difference between the main and sub-rooms. The house is set in precise arrangement, featuring closed high wall outside and layer upon layer courtyard inside, developing in a vertical or horizontal direction, forming a flexible and compact layout being closed to outside and spacious inside. For other ethnic groups, such as the Dai people living in Yunnan, every vernacular house in the village is a single-family bamboo building because in the territorial economic of the feudal system where the monogamous family is prevailed, each household must pay tax and elder children must live independently after getting married. The inner is spacious and separated simply by wooden partitions, containing only the

图 5 民居建筑群 The vernacular complex

图 6 甘肃兰州民居群
The vernacular groups in Lanzhou, Gansu Province

图 7 北京山区合院民居
The courtyard houses in the mountainous areas of Beijing

部布置以木板相间隔，也只有厅房而已，厅对外，房对内，厅兼作厨房用，这主要是由于小家庭生活方式的缘故。傣族人民虽信佛教，但无家神，寨有缅寺，宅内无供神处。也有竹楼内供神位者，则是受了汉族的影响。

2．民居的外形特征，主要表现为造型朴实、群体和谐、环境优美以及鲜明的民族特色。

我国民居建筑大多选址在溪湖林陆优美的山水自然环境之中，以取得舒适的生活条件。农村中，建筑群沿自然地势布置，或聚或散，或高或低，在城镇中则互相毗连，构成街坊。在个体造型上，根据功能需要，灵活组织空间，无论规则严整还是错落有变，它都顺乎自然，很少矫作，反映出淳朴、真实的面貌（图8）。

sitting room or living room which also serves as the kitchen. Such arrangement mainly conforms to the way of life for small families. Although the Dai people believe in Buddhism, instead of setting up shrines for Household Gods in their houses, they build Myanmar Temple in the village for worship. In some bamboo houses there are shrines for Household Gods but such practice is indeed influenced by the Han people.

2. The outward appearance of the vernacular houses, mainly manifested in plain style, overall harmony, beautiful environment, as well as distinctive ethnic characteristics

Most of the Chinese vernacular houses are located in beautiful natural landscape in order to achieve comfortable living conditions. Architectures in rural areas are deployed along the natural terrain, either gathering together or distributing in a casual manner, either setting high or low. But in cities and towns, the architectures are adjacent to each other, constituting neighborhoods. For an individual building, it is designed to organize its space in a very flexible way on the basis of functional needs. Either in regular or random arrangement, the natural design reflects the simple and real features of the vernacular houses.

China has more than 50 ethnic groups which have experienced different historic events and enjoy diversified natural living conditions. Their own production modes, customs, religious beliefs and aesthetic values, which are embodied in the vernacular houses, result in great differences, such as the thick-wall step-style flat-roof houses of the Tibetans, the arcade-style flat-roof houses of the Uygur people, the stilt houses of the Dai people, etc.

In addition to the different customs, cultural faiths and aesthetic

我国是由五十多个民族组成的国家。各个民族，由于各自经历的历史条件和聚居地区自然条件的差异，也由于民族的生产方式、生活习俗、宗教信仰和审美观念的不同，在民居建筑形式上也存在着较大的差别，如藏族的石砌厚墙台阶式平顶建筑、维吾尔族的内院拱廊式平顶建筑、傣族的独院干阑式竹楼等。

此外，我国地貌，山水丘陵多，地势高低错落，各地气候悬殊，各民族生活习俗、文化信仰、审美爱好不同，因而使得我国各民族各地区民居的外貌呈现出千姿百态、丰富多彩、灿烂而又鲜明的民族特色。

3. 民居的细部特征，包括装饰、装修、色彩、花纹、样式等，主要来源于民族的习俗、爱好、愿望和审美观念。它主要表现为丰富多彩、含蓄得体、艺术和实用的结合并寓有深刻的文化含意。

细部中，最突出的部位是大门、门窗、山墙面和某些构件装饰。由于这些构件都位于建筑外表最显目之处，它最易被人所注目，因而，长期以来就成为了民族和地方特征的重要内容（图9）。

大门在封建社会中是贫富贵贱等级的一个重要标志。在旧社会，不论贫富家庭都竭尽一切财力，为自己的住宅大门进行装饰和美化，目的是用来显示自己的门第，大门就成为了反映经济文化的象征，如北京四合院大门、内院垂花门（图10）、云南白族民居门楼（图11）、广州民居趟栊门等。人们通过大门的布置方式和形象，可以比较容易地识别这些民居究竟是属于哪个民族或哪个地区。

窗户是民居中最常见的一种建筑装修元件。

图 8 江南水乡民居
The vernacular houses in the region of rivers and lakes south of the Yangtze River

values of various ethnic groups, the hilly landscape, uneven terrain and great disparities in weather in China lead to diversified appearances of the vernacular houses in various ethnic groups which are representatives of the rich, splendid and distinctive ethnic characteristics.

3. The characteristics of vernacular houses' detail including furnishing, decoration, color, pattern, style, etc., mostly rooted in the ethnic customs, preferences, wishes and aesthetic concepts, are mainly manifested in the richness, decency, combination of art and practical use, and of profound cultural connotation.

In the detail, the most striking characteristics are of the gate, doors and windows, gable wall and some decorative elements, which easily draw the attention of the visitors as they are located in the most obvious and significant places of the architecture. Thus, over a long period, these parts have become an important content to represent the ethnic group and local characteristics.

In the feudal society, the gate (or entrance door) was an important symbol to differentiate the rich and the poor, the high and the low. In the old society, all the families, whether rich or poor, would take every means to decorate and beautify their entrance doors to show their social status, making it an economic and cultural symbol of the vernacular houses, such as the entrance door and floral-pendant of the four-in-one courtyard houses in Beijing, the gate-building of the Bai people in Yunnan, and the *tanglong* door of a vernacular house in Guangzhou. Through the set-up and image of the gate, it is relatively easy to identify which ethnic group or region the vernacular house represents.

Windows are the most common and frequently used components in a building. A window, regardless of its size, style, color, or lattice pattern,

在窗户上，无论是它的大小、式样、色彩，还是窗棂花纹、工艺，无不反映了人民的喜爱和审美心理，人们从窗户的形式上也可以判断出它是属于哪个民族或是哪个地区，如藏族的密肋饰带窗楣和梯形窗（图12），新疆维吾尔族的长条窗、尖拱窗，北方汉族四合院民居的支摘窗，广府民居的满周窗，中原地区窑洞民居的拱券窗等都是比较典型的实例。

色彩、装饰、花纹以及某些图案，由于当地民居经常使用，也成了一种独特的艺术表现手段和特征标志。江南水乡民居喜用灰瓦白粉墙，瓦片编成漏窗图案或各种植物形图案；南方民居喜用青砖墙面、陶塑脊饰；傣族民居喜用编竹装饰，如大象、槟榔树、孔雀或日、月图案装饰等。

此外，在民居中，各族人民常把自己的心愿、信仰和审美观念，把自己最希望、最喜爱的东西，用现实或象征的手法，反映到民居的装饰、花纹、色彩和式样中去，如汉族的鹤、鹿、蝙蝠、喜鹊、梅、竹、百合、灵芝、万字纹、回纹等，云南白族的莲花，傣族的大象、孔雀、槟榔树等图案。这样，就导致了各地区、各民族的民居更呈现出丰富多彩和百花争艳的民族特色。

craftwork, all mirror the taste and aesthetic psychology of the ordinary people. The form of a window is indicative of the ethnic group and region it belongs to, such as the ribbed and trapezoidal sash windows of the Tibetan, the strip and pointed arch windows of the Uygur, the removable windows in the courtyard houses of the Han people in north China, the wood-framed and colored-glass windows of the Cantonese, and the arch windows of the cave dwellings in the central plains region. All these are typical examples of the kind.

Colors, decoration, patterns and certain designs, as a result of the frequent use by the local residents, have become a unique means of artistic expression and characteristic symbol. For example, residents in the region south of the Yangtze River favor grey tile and whiting wall with tile being organized into the leaking-window pattern or various plants patterns; residents in south China are fond of black-brick wall decorated with pottery ridge; the Dai people love to use bamboos for decoration, in the patterns of elephants, nut trees, peacocks, the sun or the moon and other decorative designs.

Besides, the ordinary people of all ethnic groups often embody their own desires, beliefs, aesthetic concepts, the most desirable and favorite things, with realistic or symbolic approaches, in the residential decoration, patterns, colors, styles and other components, such as cranes, deer, bats, magpies, plums, bamboo, lilies, ganoderma, the 卐-pattern, and web design of the Han people; lotus flowers of the Bai people in Yunnan; elephants, peacocks, nut trees and other patterns of the Dai people. All these lead to the abundant and colorful ethnic characteristics of vernacular houses of various ethnic groups in different regions.

图 9 毡房入口
The entrance of the Yurt

图 10 北方民居垂花门
The floral-pendant of the vernacular houses in North China

图 11 白族民居门楼
The gate-building of the Bai people

图 12 藏族民居梯形窗
The trapezoidal sash windows of the Tibetan houses

三、中国民居的建筑艺术特征

Ⅲ. The Artistic Characteristics of the Chinese Traditional Vernacular Architecture

中国民居建筑的艺术特征主要反映在群体布局、单体建筑形象、空间组合、细部处理和装饰装修这四个方面。

(一)群体布局——和谐统一

中国传统民居布局的特点不是单座而建，而是几座合成一院，几院合成一宅，宅合成巷、街、村，再合成镇、城。民居建筑的形象不是一座建筑所能反映的，而要通过一个院落、一条巷、一条街甚至一个村落，一个墟镇，整个建筑群体才能反映出来。

南方平原地区的城镇中，因人口密集，民居建筑大多集中布局，相互毗邻，排列整齐，四周街巷围绕，表现出了严整、封闭的特点。农村中的民居，考虑到便利生产，又要节约农田和方便交通，常沿河、沿路和坡地建造，建筑有良好的朝向，并表现出一定的规则性（图13）。

在山区或丘陵地带，不少是各民族集居之地，民居建筑常沿等高线布置，有的沿山腰，有的在山脚。它的特点是自由灵活，高低错落，与自然环境协调（图14）。在汉族客家山区建造的聚居防御围楼，有单座建造，也有多座建造，有圆形，也有方形，还有两个甚至多个方形或圆形土楼交错连接的，反映在群体外貌上，体形巨大、稳重，气势豪放、粗犷。

河湖地带的民居建筑则充分利用水面，或沿河布置，或临靠水面。特别在江南地区，民宅临街背水，建筑与道路、河流的走向相适应，创造了方便生活的优美环境，充分反映了江南水乡特色（图15）。

在炎热多雨的粤中地区的村落，民居建筑密

The characteristics of the art of the Chinese traditional vernacular architecture are mainly reflected in such four aspects as group layouts, detached architecture forms, spatial combinations and detailed decorations.

i. The group layouts — harmonious and unified

The characteristics of the Chinese traditional vernacular architecture are: several houses are grouped into one yard and several yards form one homestead; several homesteads constitute a lane, a street or a village, which in turn becomes a part of a town or a city. The image of the traditional vernacular architecture can be reflected not only by a single building but also by a courtyard, a lane, a street, a village or even a marketing town.

In the towns and cities of the southern plain areas of China, with a dense population, houses are mainly arranged in groups and adjacent to each other in good order, surrounded with lanes or streets on four sides with the characteristics of neat formation and closure. For the sake of facilitating production and transportation as well as economizing farmland, houses in the countryside are commonly built along rivers, roads or on slope land with a very good orientation and certain regularities.

Quite a lot of minority nationalities are living in the mountainous or hilly country. Their houses are generally built on altitudes in line, some are half way up a mountain and some are at the foot of a hill. Such houses have the characteristics of flexible arrangement, proper distribution and being harmonious with their surroundings. The enclosed houses built for defense in the Hakka's mountainous areas are huge in contour, representing the characteristics of the people's dignity, imposing boldness and straightforwardness. Some of them are detached; some are undetached; some are round and some are tetragonal in shape; still others are interlocked with two or more round or tetragonal enclosed ones.

Vernacular houses in the river or lake areas are arranged along rivers or close to water with maximum use of the water surface. In the region south of the Yangtze River, houses face streets and with water on the back, with buildings accommodating roads and the trends of the rivers so as to create exquisite surroundings by displaying the distinguishing features of the region of rivers and lakes.

In the villages of the central part of the Guangdong Province where it is hot and rainy, the vernacular houses are in the comb-type layout because they are arranged in great density just like combs, both regular and in good order. A fishpond is commonly seen in front of a homestead with a bamboo grove behind the house. One or two banyan trees may be planted beside the rice fields. Both trees and the rice fields around them reflect the simplicity, naturalness of the houses and the rural sceneries of the farmyards of the village.

In some mountainous areas of such provinces as Fujian, Jiangxi,

集排列，像梳子一样，规则而整齐，称为梳式布局。它的前面常设鱼塘，后面种植竹林，禾埕旁又栽植一两棵大榕树，与周围稻田结合，反映出了村落民居建筑的简朴、自然和一派农家田野风光。

在南方的福建、江西、广东和山西、陕西的一些山区，由于防御的需要，民居建筑常集中组合成大型的堡寨或围楼，有单层、两层，也有多层的，或圆或方，对外封闭，对内开敞。这类建筑体形巨大、稳重而粗犷。有的土楼几个或更多地成组建造在一起，给人以强烈的印象。

在西南黔桂地区的侗族聚居的村寨，多在寨中心辟出空地修建一个高耸的鼓楼作为公共活动中心，它与村边的风雨桥和整片侗寨民居构成优美的总体天际轮廓线（图16）。

（二）单体建筑——淳朴真实

汉族地区单体传统民居一般都是单层建筑，三开间，坡屋顶，白墙灰瓦，在农村居多。也有民居建筑为两层，大多在城镇中人口密集的街巷，其形象都比较淳朴真实。有的民宅在大门、门窗或山墙、墀头部位偶然作一点装饰处理，表示美化。在山区和坡地的一些单体民居建筑，因结合地形，布局比较自由灵活，其外形简朴中带有轻巧。两层或两层以上的坡顶民居，有的每层都设外廊或向外出挑，有的则利用体形组合和挑檐，形式多样，如云贵桂湘各民族地区的吊脚楼就是实例（图17）。

四川山区和坡地地带的民居则结合地形，顺坡而建，或建筑顺着地势层层抬起，屋面也做成层层升起的形式，富有韵律感。

在民居单体建筑形象中，最显眼的部位是屋

Guangdong, Shanxi and Shaanxi, out of the needs of defense, the vernacular houses, commonly grouped into huge blockhouses or enclosed houses, which can be single-storey, two-storey or multi-storey houses, round or tetragonal in shape, externally closed but internally spacious, are huge in contour, displaying the characteristics of steadiness and roughness. Some rammed earth houses comprise several or even more blockhouses, leaving people a profound impression.

In the stockade villages inhabited by the Dong people in the regions of Guizhou and Guangxi in southwest China, an open ground is commonly marked off to build a tall drum-tower serving as the center of the public activities. The tower, the rain-proof bridge at the end of a village and the traditional vernacular houses of the Dong nationality together constitute a beautiful overall horizontal outline.

ii. The image of detached houses — simple and authentic

The detached vernacular houses in the Han people-inhabited areas, especially in the rural areas, generally one-storey, three-bay, pitched-roof ones with white walls and grey tiles are built. Two-storey houses wearing simple and authentic contours are mainly built in towns or cities that are densely populated. Some houses are casually decorated on the gate, the doors, the windows, the gable walls or the skew blocks for embellishment. Some detached vernacular houses in the mountainous areas or slope land wear a light and handy contour since they are laid out flexibly to fit the landform. Some two-storey or more-than-two-storey houses on the top of the slope land install a veranda or an overhanging eave on each storey. Some utilize form combination and projecting eaves with various forms. The overhanging houses in the provinces of Yunnan, Guizhou, Guangxi and Hunan are good cases in point.

The vernacular houses in the mountainous areas or slope land of Sichuan Province are built downhill to fit the landform or raised layer upon layer along the terrain and the roofing is also built in the same manner with the sense of rhythms.

The most conspicuous part of a detached

图 13 群体民居艺术
The art of the group-houses

图 14 坡地民居艺术
The art of the group-houses on slope land

图 15 浙江傍水民居
The vernacular houses beside water in Zhejiang Province

图 16 贵州侗寨
The Dong people's stockade

图 17 吊脚楼民居
The overhanging houses

图 18 藏族平顶建筑
The Tibetan flattop houses

图 19 塔吉克族圆帐顶
The Tajik's dome yurts

图 20 傣族歇山顶
The Dai people's shieshanding

顶。屋顶是利用当地材料，适应当地气候、地理环境条件的一种独特的审美和结构相结合的表现形式。各民族民居由于不同的自然条件、不同的材料和构造方式以及不同的生活方式所形成的不同平面，导致了不同的结构方式和外形，也就形成了各种式样的屋顶（图18、图19）。长期以来，屋顶形式已成为各地区、各民族建筑的主要特征，如汉族民居中江南的马头墙屋面、广东的镬耳墙屋面、藏族民居的平屋顶、蒙古族的圆帐顶、维吾尔族的穹隆顶、傣族的高耸歇山顶（图20）及四川民居的穿斗式披椽大屋面等都是非常典型的民居形象特征。

（三）空间组合——灵活优美

民居建筑空间分为外部和内部两部分，外部空间指环境。

民居的艺术魅力不仅是单纯依靠民居建筑本身来表现，通常还要依靠民居建筑所在的周围环境来表达。山腰茅屋因挺拔的高山才显其清秀、宁静，河畔宅居因潺潺流水才显其潇洒自在，坡地民居因建筑结合地形才显其轻巧、精致。

汉族民居的内部空间艺术主要表现在两方面：一是庭院空间，二

house is the form of its roof, which is the manifestation of the unique combination of aesthetics and structure adaptive to the conditions of the local climate and geography with local building materials. Different house plane-surfaces formed by different natural conditions, with different materials in different structural forms and different living styles of different nationalities have generated various figures of roofs. Therefore, for a long time, the styles of roofs have become the main features of the traditional vernacular houses inhabited by different nationalities in different parts of China, such as the roofs of the "horse-head" stepped gable walls of the houses inhabited by the Han people in the region south of the Yangtze River, the roofs on the pot-ear-shaped gable wall houses in Guangdong, the flat roofs of the houses inhabited by the Tibetans, the dome-shaped top of the houses inhabited by the Mongolians, the dome roof of the houses inhabited by the Uygur people, the towering xieshan-style roof of the houses inhabited by the Dai people, and the big through-jointed water drip roofing of the houses in Sichuan Province.

iii. Space composition — flexible and graceful

The spaces of the vernacular houses consist of two parts — the interior and the exterior and the latter refers to the surroundings of the houses.

Sometimes, the artistic charm of the vernacular houses is expressed by combining the architectures with their surroundings instead of merely by the architectures themselves.

The hillside thatched cottages appear delicate and tranquil because they are located straight on tall hills; riverside houses look casual and elegant because they are situated beside murmuring streams and the houses on the slope land seem light and ingenious because they are arranged to suit the landform.

The art of the interior space of the houses inhabited by the Han people is reflected in two aspects: one is the space of the courtyard and the other is the space of both the hall and interior.

1. Space of the courtyard

Most of the vernacular houses of China center their spaces in the form of courtyards or halls. Take the "four-in-one" courtyard houses of north China for example. The principal room faces south and the chambers standing on the east and west wings are arranged symmetrically along the north-to-south axis. On the south, a gallery, a tracery and the secondary gate are set up. Outside the secondary gate is the front courtyard which is long and narrow from east to west and the entrance gate is set up at the southeast corner of the courtyard. Such kind of space compositions make a distinction between the primary and the secondary ones with clear and definite partition, not only satisfying the needs of serving to pay

是厅房室内空间。

1．庭院空间

我国的民居建筑多以院落或厅堂为中心来组织空间。以北方四合院为例，院内正房坐北朝南，东、西厢房沿南北轴线对称布置。南面设游廊、花墙和二门，二门外是东西向狭长的前院，前院南面是倒座房，大门设在东南角。这种空间组合方式，主次分明，分区明确，既能满足长幼有序、内外有别的使用要求，同时，也为内院创造了安静的居住环境。

南方汉族民居庭院较小，称为天井，也有稍大的庭院，通常作为庭园。小型天井或庭院一般不栽树，因树大、杆粗、叶密，遮挡阳光且不通风。有时栽种一两株单株细木或栽竹，也有置盆景者，总的来说，以绿化为主。

中型庭院可植树一株，并辅以假山或绿化，它与建筑檐廊相连接，形成宁静、安逸的气氛。

大型庭院可置假山、池水、花木，再建楼阁、亭台或舫榭，与宅居相连，组成一组比较完整的住宅园林。园林中，引进大自然山水景色，划分大小不同的景区，再运用对景、借景手法，这样，就可以在有限的空间内，获得可居、可游、可行、可望的艺术享受。

2．厅堂内部空间

一是门窗、格扇，这是厅堂直接面向庭院天井的部位。格扇的花纹图案非常丰富，在民居中是最富于艺术表现的部位之一（图21）。

在大型民居建筑中，门窗格扇的格芯常用木条拼成方格纹、藤纹或锦纹，也有雕人物、花鸟者。有的在槛窗下半部另加罩格，作遮挡视线用。它通过大面积的图案、纹样和通透光影的对比来取

图 21　厅堂格扇
The partition board in a hall

respect for seniority and to distinguish the inside from the outside, but also creating a quiet environment for residence.

The courtyards of the vernacular houses inhabited by the Han people in the south are rather small and so are generally called small courtyards. There are slightly larger courtyards commonly used as garden courts. Generally, no tree grows in small courtyards or garden courts because the trunks of big trees are thick, leaves dense, keeping out the sunlight and affecting ventilation. Sometimes one or two single thin trees or bamboos or potted plants may be planted but generally they are mainly for afforestation.

Medium-sized courtyards may have just one tree planted, complemented with a rockery or some greenery, linked together with the eaves gallery of the architecture to generate a quiet, easy and comfortable atmosphere.

Large courtyards can have rockery, ponds, flowering wood with a multi-storied pavilion, terraces or marble boats, linked with the residence to form a group of relatively integrate private gardens. With the natural landscape, a private garden can be divided into different scenic spots. By means of scenic focal point and borrowed view in the limited space, an artistic enjoyment can be achieved in residing, touring, walking and sightseeing.

2. The space inside the hall

First, doors and windows or partition boards are built in the place where the hall leads to the small courtyard directly. The design of a partition board is rich and varied, which is one of the components rich in artistic manifestation in a vernacular architecture.

In a large vernacular architecture, the panel of a door, a window or a partition board is usually shaped with trellis design, rattan-shaped lines and brocade-like lines or with the sculpture of figures or flowers and birds. The hatch of some grid-windows is added with a covering grid for keeping out the line of sight. Through large areas of designs, lines, and the comparison between the permeable shadows, the artistic effect of the decoration and furnish is produced. In a small-sized vernacular architecture, jut window boards are commonly installed under windows, lintel eaves added to the top of windows, window bars sticking out if

图 22 厅堂梁架雕饰
The carving decorations of the beams in a hall

图 23 宅第厅堂 The hall in residence

得装饰装修艺术效果。在小型民居建筑中，往往采用窗下凸出宽窗台，窗上加楣檐，临水则凸出窗栅，或在楼层挑出栏杆等手法，来增强建筑外观上的凹凸变化和虚实对比，既符合使用要求，又增加艺术气氛。

二是在厅堂和房间内部，为了分隔空间，通常采用屏、罩、隔断等木制构件，其雕刻技术与精美图案很有特色。

三是厅堂或廊檐的梁架及其附属构件，如坨墩、梁头等，在不影响其结构性能的前提下施以雕琢，丰富了空间艺术效果，也增强了美化作用（图22）。

四是匾额、楹联，这是厅堂空间中不可缺少的内容。我国古代中原的世家望族，南迁后仍然保持着家族的历史流传，例如把堂号名称写在匾额上并悬挂在厅堂正中或门楣上，也有在厅堂明间金柱上挂对联的，称为楹联，在园林中的馆、轩、楼、阁内也大多设置有匾额、楹联，不但增加了园林的文化气息和诗意，同时也丰富了园林空间的艺术特色（图23）。

close to water, or bracket baluster on the storey to enhance the concavo-convex varieties and the void-and-solid contrast, by means of which the requirements for utilization are met and the artistic atmosphere is gained.

Second, the division of the hall and the indoor space is commonly created by adopting screens, shades or partition walls (or boards), which are characterized by carving techniques and fine designs on them.

Third, the hall's or veranda's beam mounts and their accessories such as girders, beam-ends, etc. are carved without affecting their structural property, to both enrich the spatial artistic effect and add embellishment.

Fourth, the horizontal inscribed board and couplet (written on scrolls and hung on the pillars of a hall) are indispensable for the hall. The old and well-known families, the respected and influential clans living in the Central Plains of the ancient China still maintained such family traditions after having moved to the south as hanging the name of the hall on an inscribed board on the wall or the lintel of the hall. The antithetical couplet may be hung on the golden pillars of the central bay of a hall. In addition, the horizontal inscribed boards and couplets may also be put up on the walls of a mansion, a small veranda, a storied building or a loft in a garden, which not only add the cultural tinges and poetic flavor to the garden but also enrich the artistic features of the garden's space.

Fifth, furniture and furnishing is the major means to enrich the indoor space as well as the essential tackling in a hall or housing. For example, tables, chairs, desks or tea tables are arranged in the hall for receiving guests; baldachins or shrines for ancestral tablets are set up for offering services to ancestors and the Heaven; pendent lamps or lanterns are installed for illumination at night. Wood beds, cabinets, wardrobes,

五是家具、陈设，它们是厅堂和住房中必不可少的用具，也是丰富室内空间的重要手段。如厅堂中，为接待宾客而设置了桌、椅、案、几；为祭祖祀神而设置了神案、神龛；为了夜间照明而设置了吊灯、灯笼。在住房中，则设置了供生活使用的木床、木橱、木柜和桌椅。这些用具，不但具有实用价值，而且雕作精致，富有艺术和民族特色。

（四）细部装饰——丰富多彩

1．细部处理

细部处理在民居建筑的艺术表现中占有重要的地位，其目的主要是为了强化建筑整体或局部的艺术效果。它一般都在实用部位上采用，特别是人眼可以直接看到和人手可以直接摸到的地方，更是细部处理的重点。民居建筑较多表现的部位一般都在大门、墙面、地面、房屋的梁架、柱枋、楼梯、柱础、栏杆、台阶等。

民居建筑的大门历来是户主显示其社会与经济地位的标志，因而，许多地区都对大门的式样、用料、工艺、装饰、色彩精心经营，以达到突出门第的作用。

北京地区民宅大门常用一座单独的门屋建筑，正中双扇黑色大门，门上安置铺首铁环，虽然装饰不多，但它的外貌严肃封闭，使人感到阴沉可怕。

江南地区的一些富裕大户或文人、士大夫宅第常用牌楼作为大门，门楣上加以题词，门檐做成挑檐式，并用青砖砌筑和精致的刻砖雕饰，以显示其高贵与文雅，当地称为门楼。皖南地区的古老村落，有的用牌坊作为进村的大门，如安徽歙县棠樾古村，入村前先看见牌坊群就是一例（图24）。有

tables and chairs are furnished in the bedrooms, which not only have practical values but also are decorated with exquisite carvings, full of artistic and national features.

iv. Detail decorations — rich and varied

1. Detail treatment

Detail treatment occupies an important place in the artistic expression of the vernacular architecture, the purpose of which is mainly to enhance the whole or partial artistic effect of the building. Detail treatment is usually adopted on practical locations, especially where people can see or touch with their hands directly. Locations of a vernacular architecture most displayed are usually the gate, the wall surface, the ground, the beam frames, the pillars, the staircases, the column plinths, banisters, steps, etc.

The gate of a vernacular architecture is always the symbol of the owner's social and economic status; therefore, people in different parts of the country manage the styles, materials, craft, decoration and colors of their gates very meticulously to highlight the family status.

The gate of a vernacular architecture in Beijing is usually a detached entrance hall building. A two-leafed black gate is installed in the middle of the building, on which two iron rings are fixed. Without much decoration, it still looks serious and blocked, ghastly and bloodcurdling.

Some rich, scholar or literati-and-officialdom families south of the Yangtze River commonly use a pailou (an arch over a gateway) as their gate, on the lintel of which an inscription can be seen. Projecting eaves are made of grey bricks decorated with elaborate engraving to show their nobleness and refinement. Some age-old villages in the Southern part of Anhui build memorial gateways as the entrance of the village. A typical example is the Tangyue Village in Shexian County, Anhui Province, in which one can see a memorial gateway before entering the village. The gates of the vernacular houses in some places are made in the form of recess (called the recesses gate), which effectively avoids being exposed to the sun and rain. The gate of a vernacular architecture in the central area of Guangdong Province is usually a black-varnished two-leafed one, outside which a fence gate made of some wood horizontal muntins is built, commonly called tanglong. When the gate is open, the fence gate is closed, with the purpose of both ventilation and security. Some families may install a four-leafed wood half-fence gate, which has a better effect on ventilation and keeping out the line of view. Such fence gates merge art and practicality into an organic whole with sturdy wood and beautiful designs.

The splay arch-gateway of the vernacular architecture of the Bai people is one of the most outstanding symbols of the architecture. The

的地区，一些民居大门做成凹入形，称凹斗门，既防雨，又避晒。粤中地区城镇民居则采用一种名叫“趟栊”的形式的木制门，它是在黑漆双扇大门外加上用木横条组成的一种栅门，当大门敞开时，关闭趟栊门，目的是通风，又防盗。有的还在外面装上四扇半截式木制通风栅门，更可达到通风和遮挡视线的效果。这种栅门木质坚固，图案优美，寓艺术与实用于一体。

云南白族民居，八字门楼是它最突出的标志之一。门楼平面为外开八字形，墙体用砖贴面。屋面有主次，高低有错落，墙楣、墙身、屋檐装饰艳丽。

傣族竹楼，大门为敞开式，置于竹楼二层，它由木梯直上，经廊道即见大门。大门由竹编制而成，有的还编制图案，简洁又美观。

民居建筑中，除大门外，还有二门，这是封建制度下区别内外的标志，在富裕大户中对此二门要求比较讲究和严格，如北京宅第中的垂花门。此外，还有一种洞门，多在园林中采用，其手法比较轻巧、自由，外形可做成月形、圆形、瓶形，壶形等。

墙面处理也是民居建筑细部艺术处理的重要

arch-gateway is splayed outwardly and its wall is veneered with ceramic tiles. The well-proportioned roofing comprises the primary and secondary parts while the summer, the surface and the eaves of the wall are decorated gorgeously.

The gate of the multi-storied bamboo pavilion of the Dai people is set on the second floor. One can reach it via the covered way after climbing the wood ladder from the first floor. The gate is made of bamboo braid and some gates are woven with designs, simple but artistic.

Apart from the gate, many a Chinese vernacular architecture has a second gate leading to the main court, which is the symbol to differentiate the inside from the outside under the feudal system. Some rich and influential families are very particular about the second gate, such as the floral-pendant gate of the homestead in Beijing. In addition, the artistic door opening is mainly adopted in gardens. The technique of the building is light and free and the external form may be in the form of the moon, a circular, a bottle or a kettle.

Treatment to the wall surface is also one of the most important techniques of the artistic treatment to the detail of the vernacular architectures. The contrast effect is achieved by the quality and texture of the materials employed, such as the simple but elegant effect of the white plastering walls partitioned with maroon skeleton of wood posts in the areas south of the Yangtze River and East Sichuan. In the areas surrounding Quanzhou of East Fujian, stone blocks and red bricks are alternatively used to build walls, or stone frames are inlaid on red bricks; dark and light colored bricks are used to make different designs to achieve the contrast effect of the sense of quality and texture. In the area of Chaozhou, Guangdong Province, the artistic effect is achieved with varied profiles by making layers of lines on the drooping section of the squint of the wall. On the wall surface of Yang Amiao's courtyard in Quanzhou, Fujian, there is a brick carving and on the upper part of the wall there is

图 24 安徽歙县棠樾村牌坊群
The complex of the memorial gateways In Shexian County of Anhui

图 25 庭园铺地
The paved floor of a courtyard

图 26 柱础
The column plinth

手法之一，它是依靠所用材料本身的质感来取得对比效果的。例如江南和川东地区，用栗色木柱木骨架划分白色粉刷墙面以取得素雅效果。在闽南泉州一带，则用石块和红砖穿插砌筑，或在红砖墙上镶嵌石框，用深、浅色砖砌成图案以取得质感对比的艺术效果。广东潮州地区民居建筑，在墙尖下垂带部位做成层层线条，以变化的轮廓线来取得艺术效果。福建泉州杨阿苗宅中的小院墙面上，有圆形砖雕一幅，在壁面上部又有砖雕楣檐一幅，图案优美，雕技精湛。

民居建筑中，院落地面和檐廊地面常用砖、瓦、石材铺砌，有的还在石材上施以雕刻，有的用卵石拼砌出各种图案，起到美化作用，特别在庭园、斋轩等建筑中较多采用（图25）。

柱础有承重和防潮、防水的作用，也是装饰部位之一。它常做成鼓形、瓜形、筒形、瓶形、斗形、八角形等，有做成单层者，也有做成双层的。在柱础的表面上，通常会进行雕刻，有雕花卉鸟兽者，也有雕几何图案者，其工艺和题材都有明显的地方色彩（图26）。

栏杆有木制、石造两种。木制栏杆较多用于室内，如廊下、楼梯、二楼柱廊，或楼层周围悬空部位，如楼井等，可避日晒雨淋。其处理原则是以实用为主，并与艺术相结合（图27、图28）。栏杆的实用就是安全，其艺术处理是不违背木材的结构性能，而崇尚朴实和以线条为主，适当加以浮雕或浅雕装饰。石造栏杆主要用于室外临水部位，由于水面空间不会太大，如跨水之桥，其体量也不会太大，因此，石造栏杆宜矮、宜小和坚实稳重，甚至有的园林中，临水栏杆可用条石做成，人在过桥时还可以在石造栏杆上休息，遥望池水彼岸景色，也

another brick-carving summer eaves, which are graceful in design and consummate in carving craftsmanship.

In a vernacular architecture, especially in a garden with small rooms or verandas with windows, the ground of the courtyard and eaves galleries are commonly laid with bricks, tiles or stone. Some may be decorated with carving designs on the stone, and some with various designs pieced together with cobbles for embellishment.

The column base, which is also one of the parts to be decorated, has the functions of damp-proof, water-proof. It is often made in the form of a drum, a gourd, a cylinder, a bottle, a dipper, an octagon, etc. Some column bases have just a single layer and some double layers. The surface of some column bases is commonly carved with flowers, birds or animals while that of some others is carved with geometric figures. The craft and themes are characterized with obviously local colors.

Banisters are generally made of wood or stone. The wood banisters are mainly used indoors, such as under the veranda, on the stairs, on the colonnade of the second floor, or in the parts suspended in midair such as the shaft, etc. to avoid being exposed to the sun and rain. The principle for the treatment is mainly practical combined with artistry. Banisters are practical and the artistic treatment to them gives first place to upholding simplicity with lines, decorated with adequate relief or low relief rather than go against the wood's structural behavior. The stone banisters are mainly used in outdoor sections close to water, such as a bridge over a ditch since the water is not wide. The stone banisters should be built short, small, solid and steady since the bridge may not be large. Close-water banisters in some gardens are made of stone battens, on which people can take a rest and look into the landscape on the opposite bank as an artistic enjoyment while taking a walk.

Lying on the ground, steps and platforms have the function of damp-proof. People will pay special attention to their steps when walking on them for fear of tumbling down since they are generally located in low places. Therefore, steps and platforms generally maintain their natural appearance without any treatment to the detail. If certain treatment is necessary, some lines may be simply carved on them.

2. Decoration and finishing

图 27 新疆民居廊檐
The vernacular eaves of the veranda in Xinjiang

图 28 藏居屋檐装饰
The decorations of the eaves of the Tiban houses

是一种艺术享受。

台阶、台基位于地面上，有防水防潮的作用。它虽在低处，但当人们走过时，常怕摔倒而特别注意，因此，在台阶、台基上一般不作细部处理，还它自然面貌。如需处理，也是简化，略作线条加工而已。

2. 装饰装修

装饰是附加在构件上的一种艺术处理，如屋面上的脊饰、大门装饰、外檐装饰、山墙墙面装饰、室内梁架装饰等，它们有实用价值，也不影响建筑物的使用和结构，目的是为了美化建筑物，在封建社会中，它还是显示户主地位和财富的标志。

装修也称小木作，主要指室内布置和陈设，它包括门窗、隔断、屏罩和家具等，有实用价值，也有欣赏价值。

装饰是建筑艺术表现的重要手段之一，其特征在于充分利用材料的质感和工艺特点来进行艺术加工，以达到建筑性格和美感的协调和统一。

装饰的工艺特征则是充分运用刀、锤、斧、锯等工具，直接在材料上进行构图和艺术加工，根据不同的材料采取不同的方式，从而形成不同门类的装饰艺术表现和风格。

装饰还有一个明显的特征就是意匠特征。它的艺术表现是充分运用我国传统的象征、寓意和祈望手法，将民族的哲理、伦理思想和审美意识结合起来。这种象征手法在民居装饰中较多采用，通常是用形声或形意的方式来表达。形声是利用谐音，通过某些实物形象来获得象征效果，如用莲、鱼表示连年有余，蝙蝠、梅花鹿、仙桃表示福、禄、寿等。形意则是利用直观的形表示本身的意义，如松鹤表示长寿、牡丹表示高贵、莲花表示洁净、梅竹

Decoration is a kind of added artistic treatment to the structural components, such as ridge ornament on the roofing, the decoration on the gate, the outer-eaves, the surface of the gable walls, and that on the interior beam frames, etc. It is practical, with the intent of embellishing the building without affecting the application and structure of it. It is also a symbol to show the owner's social status and wealth.

Decoration is also called small carpentry, mainly referring to the interior arrangement and furnishing, including doors, windows, partition boards and furniture, which have the values of practicality and appreciation.

Decoration is one of the important architectural means of artistic expression, the feature of which lies in making full use of the quality of the material and the techniques of the craft to carry out the artistic treatment so as to achieve the harmony and unity of the structural disposition and aesthetic perception.

The technological feature of decoration is to make full use of the tools such as knives, hammers, axes and saws to handle the material by directly making figures or artistic treatment with different methods according to different materials and thus to form different kinds of artistic expressions and styles of decoration.

Another feature of decoration is the feature of artistic conception. Its artistic expression is to make best use of the means of the traditional symbolism, the implied meaning and the wish of our country to combine the nation's philosophical theories, ethic thoughts and aesthetic awareness. Such kind of symbolism is commonly adopted in the decoration of the vernacular architecture by means of combining the form with sound or the implied meaning. By combining the form with sound, it means to take advantage of the homonym to obtain the symbolic effect through the images of some objects. For example, lotus (lotus is translated into Lian in Chinese, the sound of which means in succession) and fish (fish in Chinese is Yu, the sound of which stands for having enough) stand for "having enough and to spare in succession". Bats, sika deer and peaches stand for felicity, good salary, and longevity. By combining the form with the implied meaning, the figure of a visual object is used to indicate the content of the implied meaning. For example, pine trees and cranes stand for longevity, peonies for gentility, lotuses for cleanness and plum blossom and bamboos for exemplary conduct and nobility of character. Some people may combine the form with sound and the implied meaning at the same time to express certain ideas. For example, adding the head of a ruyi (an S-shaped ornamental object, usually made of jade, formerly a symbol of good luck) to the mouth of a bottle stands for safe and sound as one wishes. These designs and figures mainly reflect people's auspicious

图 29 维吾尔民居室内石膏装饰
The plaster decoration inside a Uygur's house

表示清高亮节。也有将形声和形意的内容合在一起使用的，如在瓶上加如意头寓为平安如意。这些图案花纹大多反映了人们的吉祥愿望，是一种具有民族特色的文化传统和观念的体现。

在民居中，装饰手法非常丰富，一般来说，它贯彻了三个原则：一是实用与艺术相结合；二是结构与审美相结合；三是综合运用其他艺术品类如绘画、雕刻、书法以及匾额、楹联等民族文化艺术的特长。这样，就增加了装饰艺术的民族特色和它的特殊感染力（图29）。

此外，装饰装修艺术表现手法还有下列特点，即构图形象上的丰富和统一，题材内容上的多样化，既可用历史人物，也可用动物、植物和花草，不拘一格。在色彩处理上，以典雅朴实为主，重点部位则稍加突出。

民居装饰在部位安排上分为室外和室内，室外以大门、屋脊、山墙和照壁为主，室内包括门窗、格扇、梁架等。以工艺类别来说，分为雕和塑两大类，包括木雕、砖雕（图30）、石雕、灰塑、陶塑、泥塑，粤东沿海一带还喜欢用嵌瓷装饰，它对防海风侵袭很有效果。

wishes as an incarnation of the cultural tradition and perception with national characteristics.

The Chinese vernacular architectures have plenty of techniques of decoration. Generally, decoration implements three principles: first, to combine practicality with artistry; second, to combine the structure with aesthetics; and finally, to comprehensively employ the special skills of other categories of national culture and art, such as painting, sculpture, calligraphy, horizontal inscribed board and couplet, etc. In this way, the national characteristics and the special artistic appeal of the decorative art is enhanced.

Besides, the techniques of the artistic expression in decoration and finishing have the following characteristics: the abundance and unity of the composition figures, the variety of themes and contents, not limited to one type, that is, historic figures, animals, plants, flowers and grass can all be employed in decoration. In the treatment of colors, the focus is put on elegance and simplicity while the major locations are highlighted more than others.

The arrangement for the locations to be decorated in a traditional vernacular architecture includes interior and exterior. The stress on exterior decoration is laid on the gate, the roof ridge, the gable walls and the screen wall while the interior decoration mainly focuses on the doors, the windows, the partition board, the beam frames and so on. The technological categories involve carving and engraving, including wood carving, brick carving, stone carving, ash sculpture, pottery sculpture, and clay sculpture. People living in the region following the line of the sea in east Guangdong are fond of using inlaid ceramic for decoration, which is effective on preventing the sea wind from hitting their houses.

Synthesizing the above-mentioned major contents and techniques of the artistic expressions of the vernacular architectures, we can sum up the general characteristics as follows:

(1) The systematization of the layout, the richness of categories and the flexibility of the combination.

(2) The Chinese vernacular architectures are prominent in adapting to local conditions, drawing on local resources and suiting certain requirements in accordance with the availability of materials under such natural conditions as the climate, geography, landforms, material, structure, etc.

This is because the architectures are built by the people, who are the designers, constructors and the users at the same time. Under the feudal system, the majority of the Chinese population was peasants and ordinary people in towns or cities, therefore their economic capabilities and requirements to the utilization of the architectures determined that the

综合上述民居建筑的主要内容和艺术表现手法来看，它的总特征可以归纳为下列几点：

(1) 布局上的规整性、类型上的丰富性和组合上的灵活性。

(2) 民居在适应气候、地理、地貌、材料、结构等自然条件下因地制宜、就地取材、因材制用的做法是非常突出的。

这是因为，民居是由人民建造的，设计者、施工者、使用者都是人民自己，三位一体。我国封建社会中人民大部分是农民和城市庶民，户主的经济能力和民居的使用要求决定民居建筑一定要走勤俭节约的道路。

(3)外形朴实，群体和谐，装饰装修丰富多彩。

我国封建社会在“藏而不露”思想的支配下，民居外形是朴实的。民居的艺术表现难于在单体建筑中表达，而只有在群体中才能得到体现。古代“中和”思想对民居建筑也有影响，例如平面布局中的中轴对称，地形处理中的前低后高、前水后山、左右环抱，在形象造型中的大小对比、稳定平衡等，都明显地表现出“和谐”的特点。由于民居使用者的经济和地位的不同，文化素质的差异，居住地域的不同等因素，其建筑的等级制度，除了规模、体型、开间等标志外，很大程度上是用装饰装修和细部来表达的。我国匠人高超熟练的工艺水平给建筑装饰装修表现提供了可能。民居装饰装修的品类齐全，构思独特，题材广泛，手法多样，它为民居建筑艺术表现增加了无限的光辉和特色。

图 30 砖雕照壁
The brick-carving screen wall

road of being diligent and thrifty in construction must be taken.

(3) The contours are simple, groups harmonious, decoration and furnishing varied and colorful.

Under the influence of being modest and not showing one's capabilities in the feudalist China, the contours of the traditional vernacular architectures were simple. It was hard to express its artistic behaviors in a detached building and so the artistic behaviors can only be incarnated in groups. The architecture was also influenced by the ideology of neutralization originated from the ancient times. For example, the central axis must be symmetric in the plane layout. In the treatment of the landforms, the front should be low while the back high, a building should face water and with hills at its back, surrounded on both right and left sides. The building must have a suitable contrast of sizes, good steadiness and balance in modeling the form of a building. All these indicate the characteristics of harmony. Apart from the symbols such as the scales, contours and bays, the hierarchical system of architectures was expressed to a large extent through decoration, furnishing and details as the economic and social status of the users varied; the cultural qualities and the places people lived in were different. The superb and proficient skills of the craftsmen made it possible to express the perception of decoration and furnishing. The decoration and furnishing have great variety of categories, unique in conceptions, extensive in themes, varied in techniques, adding boundless brilliance and features to the artistic expressions of the Chinese vernacular architectures.

四、中国民居建筑艺术表现
Ⅳ. The Artistic Expression of the Chinese Vernacular Architecture

合院民居

Courtyard houses

汉族民居中，宗法礼制、伦理思想占主导地位，反映在平面布局中，多进院落式平面为其主要模式。由于我国地域辽阔，气候差异悬殊，故平地民居可分为北方和南方两大类，北方称合院式民居，南方因人多地少，气候又湿热多雨，房屋毗连，院落较小，故称天井式民居。

北方城镇合院式民居由正房、厢房、耳房、厅房、倒座房、后罩房、大门、垂花门、抄手廊、窝角廊、穿廊、影壁、院墙等单体建筑和建筑构件所组成，其组合形式有单进院、二进院、三进院或多进院。一般宅居是沿着中轴线布置成为纵向一路，大型宅第在主轴旁带跨院，也可两路并列，也有因地形关系呈现多路交错组合的。

四合院的平面布局，正中核心庭院为南北向，一正两厢加垂花门或过厅，成为近正方形的格局。正房、厢房、倒座等房，各不相连，分布在院落周围。四合院平面组合中轴对称，院庭宽敞，有足够的阳光。正房是合院主体建筑，为主人卧室及作生活起居用。厢房位于正房的两侧，一般为儿女用房。

四合院临街一面布置倒座房，其前檐朝内院，后檐对胡同（街巷）。宅大门即位于倒座的东面。倒座不开窗，或开高窗，外貌呈现出高度封闭感。

最后一进为后罩房，作杂务院用，如临街巷，可辟后门。

北方合院式民居的特征是，前堂后寝，中轴对称，平面规整，布局严谨，内部院落相间，外观稳重封闭，反映出深刻的等级观念。

The spirit of rites and ethics occupy a leading position in the Chinese traditional vernacular architecture of the Han nationality. Planes of multiple-courtyard houses are main patterns in the plane layout. Since China is characterized by a large territory and disparities in climates, the open-ground vernacular architectures can be classified into two main types: the northern type and the southern type. The former refers to the courtyard houses while the latter small courtyard houses because the southern areas have a large population but little land and it is humid and rainy there so that houses are adjacent to each other with small courtyards.

The courtyard houses in north China consists of such detached buildings and components as the main block, wing-blocks, side blocks, the front block, posterior block, the entrance gate, the floral-pendant gate, the U-shaped corridor, short corridors (in the corners), the covered corridor (leading from one court to another), the screen wall, courtyard wall, the combination forms of which can have just one yard, two yards, three yards or multiple yards. Generally, a homestead is arranged longitudinally along the central axis. A large homestead may have a side yard along the central axis or two yards may stand side by side. Some may be arranged staggeringly according to the landform.

The plane layout of a "four-in-one courtyard" is: the central kernel courtyard is in the north-south orientation, one main block with two wing blocks added by a floral-pendant gate or a passing hall forming a pattern which is approximately a square. The main block, wing-blocks and the front block are distributed on the perimeter, not a single one is connected with anyone else. The central axis of the plane combination of the "four-in-one courtyard" is symmetric and the courtyard is spacious with abundant sunshine. The main block is the principal building, comprising the living room and bedrooms of the owner and his wife. The wing-blocks are located on both sides of the main block, generally serving as the bedrooms of the owner's children.

The front block is built on the frontage of the "four-in-one courtyard", the front eaves of which face the interior courtyard and the back eaves facing a lane (or street). The entrance gate is located on the east of the front block, which has no window at all or just a high-light window. Therefore, it assumes a look full of the sense of closure. The last block of the courtyard is the "posterior block", which is used for sundry duties such as the frontage, and the backdoor may be set up there.

The characteristics of the courtyard houses in north China are front-hall-rear-room, the central axis being symmetric, the plane being regular and tidy, the exterior being prudent and enclosed, reflecting a profound hierarchical perception.

北京四合院
The "four-in-one courtyard", Beijing

北京四合院除具有北方合院式民居的前堂后寝、中轴对称、院落相套特征外，其建筑艺术形象主要表现在厅堂、大门、二门、宅内的空间组织、庭院绿化以及室内的装饰装修上。

大门是整个四合院的起点，也是整座建筑外在形象和艺术表现的重点。因而，大门的形制、规格成为了全组建筑的等级表征，又是社会地位的标志。宅第主人为了安全与防卫需要，除加大大门的规模外，还增加了门饰、铺首、金环、门簪等附加物，门外还辅以石鼓、石兽，甚至雕花门枕石。

位于主人内院之前的出入门口称为垂花门，也是大宅的二门，一般做得比较华丽，是四合院内的装饰重点。

四合院的院落空间组织是充分运用廊、墙，或间隔，或相连，形成穿廊、花墙，院内则栽植少量树木或置盆景，呈现出幽静、安逸、舒适的气氛。

Apart from the north courtyards' common characteristics of front-hall-rear-room, the central axis being symmetric, courtyards being interlinked with each other, the artistic image in the architectures of the Beijing "four-in-one courtyard" is mainly reflected in the hall, the entrance gate, the secondary gate, the spatial texture inside the homestead, the greening of the court and the interior decoration and finishing.

The main entrance is the starting point of a whole "four-in-one courtyard", and also the focal point of the artistic expression of the whole architecture's external form. Therefore, the shape and structure, and the classification of the main entrance became the hierarchical characterization of the whole set of buildings, and the symbol of the family's social status as well. Apart from enlarging the scale of the main entrance, the owner of a homestead adds such auxiliaries as the decoration on doors, the Pushou (the iron plates fixed on the doors to keep the knockers), the golden door-knockers, the door clasps, complemented with the stone drums, stone animals and even the engraved (door stone/crossties) outside the main entrance.

The entrance located in front of the interior courtyard is called the floral-pendant gate, also the secondary gate, which is the focal point of decoration in the "four-in-one courtyard" and generally decorated magnificently.

The space composition of the "four-in-one courtyard" makes full use of the veranda, the wall to form a passing corridor by means of spacing or connecting. A small number of trees or miniature trees and rockery may be planted in the courtyard so as to create an atmosphere of quietness, easiness and comfort.

北京四合院建筑 The four-in-one courtyard houses in Beijing

北京四合院正房檐廊 The main house's veranda eaves of the four-in-one courtyard in Beijing

垂花门 The floral-pendant

二门内照壁 The screen wall inside the second gate

四合院抄手廊与小院
The U-shaped corridor and the small courtyard of the four-in-one courtyard

四合院二门　The second gate of the four-in-one courtyard

天津石家大院

The Residence of the Shi Family, Tianjin

石家大院位于天津杨柳青镇，是清代津门八大家之一石万程第四子石元士的住宅，堂名“尊美堂”。该宅始建于1875年，南北长96米，东西长62米，占地6080平方米，建筑面积2945平方米，是天津典型的四合院住宅建筑。

宅居的总平面由东院、西院和跨院组成。

东院为住宅院落，由前后三套四合院组成，各院正房为五开间，中间的明间为穿堂。东、西厢房均为三开间。西院是厅堂院落，由两座回廊院连接而成，是石家大院中规模最宏大、装修最精美的一组建筑。

南面由两座青瓦硬山顶和卷棚顶前廊构成，前面为花厅，在接待贵宾及宴会时用，后面为戏楼。

Located in Yangliuqing Town of Tianjin City, the residence of the Shi Family is the house of Shi Yuan-shi, the fourth child of Shi Wan-cheng, one of the eight grand families in Tianjin in the Qing Dynasty. Renowned as “Zun Mei Tang (Respect & Beauty Hall)”, the house originally built in 1875 is a typical courtyard residence in Tianjin, with a length of 96 meters from north to south, a width of 62 meters from east to west, covering an area of 6080 square meters and a floor area of 2945 square meters.

The general layout of the house is composed of the east yard, the west yard and the side yard.

The east yard is for living, consisting of three sets of courtyards. The principal room of each courtyard is in five-room breadth with a passage way in the middle. The wing-rooms in the east and the west are of three-room breadth. The west yard is a hall courtyard connected by two winding corridor yards, which is the most magnificent and the most beautifully decorated complex in the residence.

The south part of the house is formed by two front corridors of blue tile, hard top and paraboloid top. The front part serves as a hall for guests reception and banquets and the rear part is a theater.

垂花门 The floral-pendant

后院门洞 The entrance of the back yard

石家大院厅堂 The hall of the Residence of the Shi Family

内院 The interior yard

内蒙古呼和浩特恪靖公主府

The Mansion of Princess Kejing, Hohhot, Inner Mongolia

固伦恪靖公主府位于内蒙古自治区呼和浩特市，即过去的归化城城北。固伦恪靖公主下嫁之际，正是喀尔喀蒙古时局动荡之时，而归化城局势稳定、商业繁荣，是大漠南北重要的物资集散地，加之距京城较近，成为建府首选。

公主府始建于康熙三十六年（1697年），占地1.8公顷，建筑面积4800平方米，共计房舍69间，其规模超过当时的归化城都统衙门。

公主府由皇家督造，依朝廷工部大式营建，以大青山为屏，札达海河与艾不盖河环抱。其风格与明末清初京畿地区的王府相似，采用中国古代建筑体系中传统的中轴对称建筑格局，大面积夯筑地基，硬山式建筑特征。

府第分四进五重院落，前有影壁御道，后有花园马场、府门、仪门、静宜堂、寝宫、配房、厢房、后罩房依次分布，是目前保留最完整的清代公主府第，也是最为典型的清代四合院建筑组群。

The Mansion of Gu Lun Princess Ke Jing is located in Hohhot, Inner Mongolia, in the north of Guihua City in the old days. It was a time of turmoil in Khalkha-Mongolia when Gu Lun Princess Ke Jing got married, but Guihua City then enjoyed a time of stability and commercial prosperity. As an important north-south material distribution center in the desert and with the geographical advantage of being close to Beijing, Guihua City became the first choice to build the mansion.

The Princess Mansion was built in the 36th year of Emperor Kangxi of the Qing Dynasty (Year 1697), covering an area of 1.8 hectares and a floor area of 4800 square meters, a total of 69 houses. It was larger than that of the Governor's Mansion in Guihua City at that time.

As an imperial residence, the mansion was constructed according to the standard prescribed by the Ministry of Construction of the Qing Dynasty, taking the great green mountain as the screen and lying in the arms of Zhadahai River and Aibugai River. Its style is very similar to the palaces in the capital city of Beijing which were built at the end of the Ming Dynasty and the beginning of the Qing Dynasty, adopting the traditional central axis of symmetry in architectural pattern in architectural system in the ancient China, constructing a large area of foundation and boasting hard-mountain features.

The mansion is a four-hall and five-yard residence, with a screen wall and a road for imperial carriage in the front and in the rear, garden, racecourse, house gate, ceremonial gate, Providence Hall, living room, wing rooms, backside house are arranged in hierarchical order. Till today, it is not only the most well-preserved princess mansion in the Qing Dynasty, but also the most typical courtyard building in that period.

寝宫垂花门 The floral-pendant of the Resting Palace

寝宫内院 The interior yard of the Resting Palace

公主府大门 The main gate of the Mansion of Princess

静宜堂 The Providence Hall

静宜堂厢房 The Wing-rooms of the providence Hall

后罩房 The backside house

寝宫主房 The main house of the Resting Palace

寝宫内院入门

The entrance of the interior yard of the Resting Palace

山东曲阜孔府

The Confucius Mansion, QuFu, Shandong Province

阙里圣人家——孔府位于山东曲阜东华门大街。孔府坐北朝南，横向分东、西、中三路布置，共有房屋四百余间，占地240多亩。中路是孔府的主体部分，前后九进院落，前为官署，后为内宅。

孔府大门三启六扇，二门也是三启六扇。二门院内，柏桧参天、灌木繁茂。过二门后，经重光门，到大堂。大堂后为二堂，有通廊相连，呈工字形平面。堂之左右为厅。三堂六厅之后便是内宅，充分体现中国古代前堂后寝制的布局模式。

内院分前上房，前堂楼、后堂楼、佛堂楼等，为家眷住房。堂楼为两层，楼上有回廊、栏杆。内宅建筑简洁大方，浑厚实用。中路最后是后花园。

Queli Saint's Residence — the Confucius Mansion is located in Donghuamen Street of Qufu, Shandong Province. The Mansion faces south and is divided into three layouts: east, west and center, having more than four hundred houses and covering an area of 240-mu. The central part is the main body of the architecture, consisting of nine courtyards with government offices in the front and residential quarters in the back.

The gateway of the Confucius Mansion has three doors and six door panels, and so is the second gateway. Within the second gateway, there grow high cypress trees and lush bushes. After the second gateway and through the Special Glory Gate, comes to the largest hall, which is followed by the second largest hall. These two halls are connected through a veranda, displaying a 工-shaped plane. At the left and right sides of halls are chambers. After three halls and six chambers is the inner living quarters. Its layout fully embodies the architectural features in the ancient China, government offices in the front and inner quarters in the back.

The inner quarters are living places for family members, including front main room, front chamber, back chamber and family chamber for worshipping Buddha, etc. Chambers are of two-floor architecture, with winding corridors and railings on the second floor. The design for the inner quarters is simple, elegant, grandeur and practical. At the end of the central part is the back garden.

孔府大门入口 The entrance of the gateway of the Confucius Mansion

孔府大门 The gateway of the Confucius Mansion

孔府内院 The interior yard of the Confucius Mansion

仪门装饰　The decoration of the ceremonial gate

孔府仪门 The ceremonial gate of the Confucius Mansion

内院建筑 The architecture of the interior yard

孔府内院 The interior yard of the Confucius Mansion

厅堂陈设 The furnish of the hall

后院入口 The entrance of the back yard

后花园 The back garden

宅居檐廊 The eave corridor of the mansion

孔府后院旁门
The side door of the back courtyard of the Confucius Mansion

后院堂楼 The major building of the back yard

山东栖霞牟氏庄园

The Mou-Clan Manor, Qixia County, Shandong Province

牟氏庄园坐落于栖霞县城北古镇都村。庄园坐北朝南，东西长158米，南北长148米，分三组六院，共建厅堂楼房480余间，自清雍正年间起至民国二十四年止，历经五代人才建成。从整体上看，属于套院式布局，即大院套小院，小院为四合院、三合院或二合院。纵向重重院落相套，横向以更道相通。主体建筑物南北纵向排列，形成一条明显的轴线。小院和次要房间大体左右匀称。空间主次分明，秩序井然。

牟氏庄园各宅院内除堂屋、宅居外，还有书斋、账房、粮仓、酒坊、碾坊、大伙房等，一应俱全，体现了小农经济的思想，自给自足，万事不求人。

牟氏庄园是山东胶东民居的典型代表。

The Mou-Clan Manor is located in Du Village, an ancient town in the north of Qixia County. The Manor, facing south, 158 meters long from east to west, 148 meters wide from north to south, is divided into three groups and six courtyards, with more than 480 halls and main rooms. Its construction began in the reign of Emperor Yongzheng in the Qing Dynasty and was completed in the 24th year of the Republic of China through the efforts of five generations. Generally, it has a multi-courtyard layout, that is, large courtyard contains small courtyard which is usually enclosed in four directions, three directions and two directions. Courtyards set in one another in the longitudinal direction and are connected through small lanes in the latitudinal direction. The main buildings are arranged in south-north orientation so that an obvious axis is formed. Small courtyards and secondary rooms are planned in a symmetrical manner. Such arrangement creates a sense of order and focus with clear priorities.

In addition to the halls and living rooms, the Mou-Clan Manor also houses a study, an accountant's office, a granary, a wine shop, a grinding mill, a big kitchen and so on. With everything being readily available, the architecture reflects the thinking of small farmers, being self-sufficient and self-dependent.

The Mou-Clan Manor is a typical representative of the vernacular houses in the areas of Jiaodong Peninsula, Shandong Province.

宅居主楼 The main building of the manor

厅堂檐廊 The eave veranda of the hall

庄园大门 The Gate of the manor

山墙与烟囱装饰 The gable wall and the decoration of the chimney

内巷 The interior lane

戏台 The stage

内院一角 A corner of the interior yard

山东章丘朱家峪

Zhu Jiayu Village, Zhangqiu City, Shandong Province

章丘市官庄乡朱家峪村，是中国北方地区典型的山村型古村落。明洪武二年（1369年），朱氏进村，因系国姓，故名朱家峪。

自明代以来，虽经六百余年沧桑，朱家峪仍较完整地保存了原来的祠庙、楼阁、宅居、石桥、古道、井泉和古哨等村落建筑空间格局。

朱家峪为梯形聚落，上下盘道，高低参差，错落有致。村子三面环山，北面依平原。朱家峪有大小古建筑近两百处，各种石桥二十余座，井泉二十余处，庙宇十余座，有被誉为"世界立交桥原型"的康熙双桥，有被称为"古代交通先驱"的双轨古道。双轨古道始建于明代，清朝时进行重修。还有文昌阁、关帝庙、朱氏家祠等，人文、自然景观数不胜数，被誉为"齐鲁第一古村"。

Zhujiayu Village of Guanzhuang Town, Zhangqiu City is a typical historic mountainous village in north China. In the 2nd year during Emperor Hongwu's reign of the Ming Dynasty (1369), the Zhu family came to settle down in the village. Since Zhu was the imperial surname, the village was given the name Zhujiayu.

Over 600 years of vicissitudes since the Ming Dynasty, Zhujiayu still keeps intact its original architectural-spatial layout, including the shrines, the pavilions, the houses, the stone bridges, the ancient roads, the wells and springs, the watching-towers, etc.

Zhujiayu is planned in trapezoid shape, paths winding up and down the mountains and buildings being deployed in different heights, setting with each other. The village is surrounded with mountains in three directions and borders on a plain in the north. There are nearly 200 ancient buildings in different sizes, over 20 stone bridges in different kinds, more than 20 wells and springs and 10-odd temples. Of them, the Kangxi Double-Bridge is known as "the prototype of the world overpass"; the two-track ancient road is reputed as "the ancient transportation pioneer", which was built in the Ming Dynasty and renovated in the Qing Dynasty. Besides, the cultural and natural scenes, such as the Wenchang Pavilion, the Guandi Temple, the Zhu-Clan Ancestral hall, to name only a few, have won the village "the first ancient village in Shandong".

民居大门装饰　The decoration of the gate of a vernacular house

朱氏家祠大门　The gate of the Zhu Clan's Ancestral Hall

拾级而上的民居　The houses ascending the stairs

村口门楼 The gate-building at the mouth of the village

关帝小庙 The Guandi God's Temple

村落小巷 The lanes of the village

古村街道 The streets of the historic village

民居入口 The gate of the vernacular house

山西襄汾丁村民居

The vernacular architecture of Dingcun Village, Xiangfen County, Shanxi Province

丁村位于襄汾县城关镇，现存明清民宅40余座，分三大部分，为北院、中院、南院。

北院位于村的东北部，以明代建筑为主；中院位于村的偏东部，以清代中早期建筑为主；南院在村的西南部，以清代晚期建筑为主。由于建造年代不同，民居的布局、建筑风格也各不同。

明代民宅多以“口”字形四合院为主。大门设在东南角，院内宽敞舒适，整组建筑较低，屋顶平缓，檐枋通施彩绘，雕刻简单古朴。清代中早期民居比明代的大，布局多为“日”字形，前后两进院，大门设在正南，院庭窄而长，硬山顶，屋顶陡起。一般不施彩画，只用红、黑色油饰檐枋和柱子。清代晚期民宅较中早期更高大，用材也更细，布局趋于组合形。其明显特点是：正房楼上楼下通用花棂装修，气魄宏大。

Dingcun Village is situated at the area just outside the town gate of Xiangfen City, Shanxi Province. The existing 40-odd vernacular houses built during the Ming and Qing Dynasties are divided into three sections: the north courtyard, the middle courtyard and the south courtyard.

The north courtyard is situated in the northeast of the village, mainly built during the Ming Dynasty. The middle courtyard is situated in the east of the village, mainly built during the early and the medium years of the Qing Dynasty. The south courtyard is situated in the southwest of the village, mainly built during the late years of the Qing Dynasty. The layouts and the architectural styles vary because they were built in different years.

The vernacular houses built during the Ming Dynasty are mainly shaped as four-sided “four-in-one courtyard”. With the main entrance set up at the southeast corner, the courtyard is spacious and comfortable. The houses are built low, with gently sloping roofs, colored decorations on the eaves and the square columns and simple and unsophisticated carvings. The houses built during the early and the medium years of the Qing Dynasty are larger than those built during the Ming Dynasty, mainly laid out in the shape of 日, with two yards: the foreyard and the backyard. The main entrance is set up at the due south of the courtyard, which is narrow and long with hard gable tops and steep roofs.

内院 The interior yard

丁村民居外观 The outward appearance of the vernacular architecture of Dingcun Village

照壁 The screen wall

三间牌坊 Three-bay memorial gateways

民居外檐装饰 The decoration of the outer-eaves of the vernacular house

民居后楼入口
The entrance of the building on the back of the vernacular house

柱础装饰 The decoration of the column plinth

花墙院门 The gate of the tracery-wall yard

山西祁县
乔家大院
The Qiaojia Compound, Qixian County, Shanxi Province

乔家大院始建于清乾隆年间，是一座城堡式建筑。大院四周是全封闭的高墙，上层是女儿墙和垛口，还有更楼、眺阁，很有气势。宅第大门坐西向东，大门对面是砖雕百寿图照壁。大门内是一条石铺地甬道，甬道尽头是祠堂。

整个宅院分为六个大院，分置两边，北面三个大院，布局是本地典型的"里五外三穿心楼院"，即里院正房、厢房都是五间，外院厢房却是三间，里外院之间有穿心厅相连。南面的三个大院都是二进四合院。六个大院，各由三五个小院组成，院中有院，院中套院。各院房顶有走道相通，便于夜间巡更护院。

乔家大院设计精巧，建筑考究、规范而富有变化，既有整体美感，局部又各具特色，无论是砖雕、木雕、彩绘，还是建筑上的斗栱、飞檐、门窗等，均形式各异，变化多端。

As a castle-typed architecture, the Qiaojia Compound was originally built during Emperor Qianlong's reign of the Qing Dynasty. With the enclosed high walls all around the courtyard, parapet on top, and with the watchtower and the balcony, the courtyard looks momentum. The entrance of the compound faces east and the opposite side of it is a screen wall with brick-carvings of the scroll consisting of hundred forms of the character for longevity. Inside the main entrance is a stone paved path leading to the clan hall at the end of the courtyard.

The whole compound consists of six courtyards, three of them on each side of the road. The three courtyards on the north are typical of the local character of "the hallway courtyard", i.e. there are five principal rooms and five wing-rooms in the interior courtyard respectively but only three wing-rooms in the exterior courtyard, with a hallway connecting the two yards. The three courtyards on the south are all "double four-in-one courtyards". Each of the six courtyards consists of three-to-five small yards, with small yards in a larger one and one yard linked with another. The top of all the houses in all the yards have footpaths convenient for keeping night watch against outside attacks.

Ingeniously designed and beautifully constructed with specifications and variety, the Qiaojia Compound has the aesthetic perception as a whole while each part has some unique features of its own. Whether the brick-carvings, the wood-carvings or the bucket arches, upturned eaves, doors and windows, etc. are all of different forms and the most changeful.

乔家大院内街 The interior streets of the Qiaojia Compound

入口门楼 The entrance of the gate-building

泰山石敢当照壁 The screen wall of the Shigandang, Mount Tai

屋顶装饰 The roofing decoration

民居院落 The vernacular courtyard

内院堂楼 The main building of the interior yard

山西祁县
渠家大院
The Residence of the Qu Family, Qixian County, Shanxi Province

渠家大院位于山西祁县县城东大街东段，始建于清乾隆年间，由16个四合院组成，共187间房屋，为全国罕见的五进式穿堂院。大院外观为城堡式，墙高10余米，墙头有垛口式女儿墙。

第一院为石雕栏杆院，上刻鱼虫花草，工艺精湛。大院右侧为五进院，庭院森森。主院为二进式牌楼院，一座10余米高的十一踩木制牌楼高耸其间，设计精巧，工艺精良。

主院为戏台院，主院两侧有一条青砖通道，将南北六个院子一分为二。北面两个统楼院，南面四个相通的四合院。整个院落，青石奠基，水磨砖墙，屋顶多样，形式各异。院内建筑布局合理，楼房错落有致，院落主次分明，屋内外彩绘华丽，木雕、石雕、砖雕玲珑剔透，整个大院建筑精致高雅，富丽堂皇。

The residence of the Qu Family is located in the eastern section of the East Street of Qixian County, Shanxi Province. Built during the reign of Emperor Qianlong of the Qing Dynasty, the residence is composed of 16 courtyards with a total of 187 rooms, a rare five-hall hallway courtyard in the nation. From the outside, the residence looks like a castle, being guarded by 10-meter high walls with parapet walls in the crenel-style.

The first courtyard features in the stone railings carved with fish, insects and flowers, reflecting an exquisite workmanship. Its right side is the five-hall compound with an exuberant garden. The main courtyard is a two-hall arch-building compound, a wooden archway over 10-meter-high with 11 pillars towering in the place, showing delicate design and sophisticated technology.

The main courtyard is designed as a stage courtyard and two brick roads on both sides divide the six yards in north-south direction into two parts. On the north, there are two two-story buildings and on its south are four interlinked courtyards. The entire courtyard, with bluestone foundation, water mill bricks, roofs in varied forms, presents a charming effect. Architectures are arranged in a reasonable layout; rooms are placed in a random but nice way; courtyards are constructed in a prioritized manner; inside and outside the buildings are painted with ornate pictures and decorated with exquisitely carved woods, stones and bricks. The whole architecture is artistic, elegant and magnificent.

内院隔墙门洞 The wall gateway of the interior yard

院落空间 The space of the courtyard

门楼雕饰 The carving decorations of gate-building

宅居堂楼 The main building of residence

山西灵石王家大院

The Residence of the Wang Family, Lingshi County, Shanxi Province

王家大院位于山西省晋中市灵石县的静升村，由东宅院高家崖和西宅院红门堡以及宜安院、孝义祠、当铺院、王氏宗祠、戏台院、佣工院等共同组成，总计有大小院落123座、各种房屋1118间，总面积约25万平方米，是国内现存规模最大的古代民居建筑群。

西宅院红门堡创建于清高宗乾隆四年至五十八年（1739–1793年），东宅院高家崖创建于清仁宗嘉庆六年至十六年（1801－1811年），东宅院与西宅院两大建筑群隔沟对峙，一桥相连，相辅相成，相映成趣，浑然一体，气势宏伟。建筑组群依山而建，再加上匠心独运的砖、木、石雕装饰，内容丰富美观。

The residence of the Wang Family is located in Jingsheng Village, Lingshi County, Jinzhong City, Shanxi Province. It is composed of the east courtyard of the Gaojiaya compound and the west courtyard of the Red Door Fort, as well as the Yi'an yard, the Xiaoyi Temple, the Dangpu yard, the Wang-Clan Ancestral Hall, the stage yard, the servants' yard, a total of 123 large-and-small-sized courtyards and 1118 rooms. Covering a total area of 250000 square meters, the residence is the largest existing ancient vernacular architectural complex in China.

The west courtyard of the Red Door Fort was constructed from the 4th year to the 58th year during the reign of Emperor Qianlong's reign of the Qing Dynasty (1739-1793); while the east courtyard of the Gaojiaya compound was built from the 6th year to the 16th year during the reign of Emperor Jiaqing of the Qing Dynasty (1801-1811). Two architectural complexes standing face to face with each other across the ditch are connected by a bridge, creating a unified and grand scene. Constructed against the mountainside, the architectural complexes are enhanced in their beauty through specially-selected bricks, wood and stone carvings, presenting a rich content and appearance.

二门牌坊 The memorial gateway on the second gate

民居院落 The vernacular courtyard

二楼檐廊 The eave veranda on the second floor

窑居室内 The interior of the cave dwelling

厅堂外檐 The outer-eaves of the hall

院门装饰 The decoration of the gate

檐柱装饰 The decoration of the eave's columns

石雕牌坊 The stone-carving memorial gateway

墀头砖雕 The brick-carving decoration of the steps of the gate

照壁雕饰 The stone carving of the screen wall

内院大门 The gate of the interior yard

窑居外檐木雕装饰
The wood carving decoration of the outer-eaves of the cave dwelling

牌坊抱鼓石雕饰
The drum stone carving decorations of the memorial gateways

民居外檐门罩木雕
The wood carving on the French door′s outer—eaves of the vernacular house

照壁雕饰
The carving decorations of the screen wall

陕西韩城
党家村民居
The vernacular architecture of Dangjia Village, Hancheng County, Shaanxi Province

党家村始建于元至顺二年（1331年），此后有三次大规模建设。

村坐落在黄土塬间沟谷中，村南有泌水绕行，依塬傍水，既可屏障寒风袭击，获取良好日照，又可依据地势组织排水。虽处黄土地区，但风尘较少。

党家村风景优美，瓦屋千宇，绿树成荫，村落、塔碑、楼祠与民居辉映。寨墙和险峻崖身高出村址30多米，极为壮观。村中昂首挺立的望楼、塔碑构成了村落优美的天际轮廓线。

党家村巷道路面石墁，丁字形，巷不对巷，门不对门，而且大门不冲巷口。进村的巷口有防盗哨门25处，极有利于防御。

党家村宅院的门楼高大，门楣上有精美的木雕和显示地位的匾额，门两侧的墀头和屋脊上都有精致的砖雕装饰。

Dangjia Village was founded in the second year of the Zhishun Emperor's reign in the Yuan Dynasty (1331) with three times of large-scaled construction afterward.

The village is situated in the valley of the loess plateau, with the Mishui River passing around the south of it. Situated at the foot of the plateau and beside the river, the village is not only protected against cold wind and gets adequate sunshine, but also is able to drain water in accordance with the topography. That's why it is not very windy and dusty in the village though situated in the loess area.

Dangjia Village boasts itself for its scenic beauty and tile-roofed houses. The trees make a pleasant shade. The village, the stone tablets, the clan halls and the vernacular houses add radiance and beauty to each other. The bulwark of the village and the precipitous cliff are more than 30 meters higher than the ground of it, looking extremely magnificent. The upright-standing belvedere and the stone tablets in the village form the beautiful outline of the horizon.

The pavement of the village streets is made of stone blocks. With the T-shape, the lanes and the gates of the village do not face each other and the gates do not face the entrances of the alleys. There are 25 sentry posts guarding against theft at the entrance to the village, contributing greatly to the defense.

The gate-building of Dangjia Village are tall and big, with exquisite wood carvings and horizontal inscribed boards on the lintels and the steps on both sides of the gate and the ridges of the houses all have elegant brick-carving decorations.

民居内院 The interior yard of the vernacular house

村落鸟瞰 A bird's eye-view of the village

内院门楼 The gate-building of the interior yard

村落街道 A street in the village

民居大门
The gate of the vernacular house

门楼雕饰
The carving decorations of the gate-building

河南安阳
马氏庄园
The Ma-Clan Manor, Anyang City, Henan Province

马氏庄园位于河南安阳市蒋村镇西蒋村，为清末兵部侍郎、都察院右都副使、广西巡抚马丕瑶的大型府第。建筑群分南区、中区、北区，共6组，22个院落。建筑面积5000平方米，占地面积20000平方米。门、厢房、堂、廊、楼共计308间，规模宏伟，系中原最大的清代官僚豪宅。

庄园保存基本完整。北区为其祖居地，建筑坐南朝北，是一座较大的两进四合院；中区建于光绪年间，规模最大，东西并排四路院落组群，即沿门前东西大道并列四路院落，均坐北朝南。当地民众习惯把传统建筑纵向中轴线统领的宅院称为“路”。西边三路院落以中路为主院，两边为次院，属一主两次并列的深宅大院。这三个院落的格局相近，全部由四进院落组成，东、西院面阔略小于主院。最东面的二进四合院为马氏家庙和家族学堂。南区建于民国初期，其格局、风格也与中区建筑群相近。

The Ma-Clan Manor is located in Xijiang Village, Jiangcun Town, Anyang City, Henan Province. It was a large-scale residence of Ma Pi-yao, Deputy Minister of Defense in the late Qing Dynasty, Deputy Surveillance Commissioner of the Ministry of Supervision and Governor of Guangxi. The architectural complex is divided into south region, central region and north region, with 6 groups and 22 yards. It covers an area of 20000 square meters with a floor area of 5000 square meters. As the largest official residence in the Qing Dynasty in Central China, it boasts 308 doors, wing-rooms, halls, corridors and buildings.

The original manor has been largely preserved. The north region, a large two-hall courtyard, is the ancestral residence and all the buildings are looking northwards. The central region, the largest compound in the manor, was built during the reign of Emperor Guangxu of the Qing Dynasty. It is composed of four courtyards sitting side by side from east to west and along the east-west orientation road in front of the door and all the four courtyards face south. The locals used to call the yard “route” which lies along the vertical axis in the traditional architecture. The west region is a courtyard with one main yard in the middle and two secondary yards one both sides. The layout of each courtyard is similar to each other and each is composed of a four-hall yard. The breadth of both the east and west yards is smaller than that of the main yard. The two-hall courtyard on the far east side is the ancestral hall and the family school of the Ma clan. The south region was built in the early period of the Republic of China and its pattern and style are similar to that of those buildings in the central region.

院落 The courtyard

堂楼 The main building

南区院落 The courtyard of the south section

北区入口大门
The entrance of the gate in the north section

北区二门 The second gate of the north section

马氏庄园南区院落中轴大门
The vertical-axis gate of the courtyard in the south region of the Ma-Clan Manor

西路侧院 The side yard in the west

二楼檐廊 The eave veranda on the second floor

过厅陈设 The furnish of the passage hall

河南博爱寨卜昌村

Zhaibuchang Village, Boai County, Henan Province

寨卜昌是一个古老的村庄，其地理位置优越，北依太行，南濒沁水。据传，周武王兴兵伐纣时，八百诸侯会孟津，向东进军来到这太行之阳、沁水之北的地方，姜太公观天象、查地理，占卦显示“昌”字，以示此地为风水宝地，日后必定繁荣昌盛，遂曰“卜昌”。

此地先民就本族姓加“卜昌”二字定村名，有乔卜昌、侯卜昌、刘卜昌等；也有在本族姓前再加自己所从事的行业的，如药王卜昌、油王卜昌等。历史上共八个卜昌村，其中药王卜昌、油王卜昌、乔卜昌三个村庄相邻互为一路之隔，为防匪患而统一围合寨墙，后于1946年更名为“寨卜昌”。

寨卜昌村现有古民居130座，400多间房屋，最早的建于清嘉庆年间，最晚是光绪年间所建。临街大门的位置较为固定，绝大多数院落大门是严格按文王八卦位布置的，设于东南方。宅院格局进深多为两进，坐北朝南的建筑宅院大多是东为主院，西为跨院。

Zhaibuchang is a historic village with a privileged location, facing the Taihang Mountains in the north and the Qinshui River in the south. The legend has it that when King Wu of the Zhou Dynasty was fighting against the brutal King Zhou of the Shang Dynasty, eight hundred lords met in Mengjin, marched their army eastwards to this place which is located in the south of the Taihang Mountains and north of the Qinshui River. Here, Jiang Taigong, the wise man, after studying the astrology and inspecting the geography, took a divination of "Chang (prosperity)", which indicated that this place was a lucky place and would become rich and prosperous in the future. Hence the place was called "Bu-Chang".

Local ancestors named their villages by adding "Bu Chang" to their surnames, like the villages of Qiao Buchang, Hou Buchang, Liu Buchang, etc. Some even placed their trades before their clan names, such as Medicine Wang Buchang, Oil Wang Buchang and so on. In history, there are eight Bu Chang villages, of which Medicine Wang Buchang, Oil Wang Buchang and Qiao Buchang are bordering each other on just one path. In order to prevent banditry, these three villages were entrenched with unified wall, and thus was renamed "Zhai Buchang" in 1946.

Today, Zhai Bu Chang Village has 130 vernaculars with 400 houses. The oldest was built during the reign of Emperor Jiaqing of the Qing Dynasty. Mostly, the location of a gate to street is fixed since the vast majority of the courtyard entrances were deployed in the south-east direction, strictly following the order of the Eight Diagrams invented by King Wen of the Zhou Dynasty. The architectural layout is usually a two-hall yard and most buildings which face south have the main yard in the east side and wing-yard in the west side.

寨卜昌村民居 The vernacular house of the Zhaipuchang Village

宅居主院 The main courtyard of the residence

内院庭园 The interior yard

入口大门 The entrance of the gate

大门抱鼓石 The drum stone on the gate

屋顶装饰 The roofing decoration

外墙拴马孔 The hole for tying a house on the exposed walls

甘肃
天水南宅子
Nan Residence, Tianshui City, Gansu Province

天水南宅子始建于明万历年间，为明代山西按察司副使胡来缙居所。南宅子以南北为纵轴，并列三组院落。由东至西分别为主院，次院以及附属院落，功能明确。主院为主人祭祀、生活起居场所；次院为读书习文、休闲娱乐场所；附属院落则是堆放杂物和仆人居住的地方。

南宅子大门位于院落平面中间，面阔三间，门内为一小天井，小天井正面是影壁墙，东西各设一道垂花门，分别为主院和杂院入口。

主院第一进——桂馥院，正厅五开间，明间和两次间是会客场所，两侧稍间分别用墙隔开，为主人或年长者居住。第二进——槐荫院的正房与东西厢房均为三开间，正房为祭祀祖先之地，两侧厢房为小姐闺房。次院的第一进院落为书院，包括正厅、倒座、过门及佛堂，均面阔三间。穿过书院正房即到棋院，棋院正房也是三开间。附属院落与主院、次院并不直接相连，而是由大门东侧垂花门进入，与次院之间有一条巷道相隔。第一进院落为杂院，后面为仆人居住的凌霄院。南宅子以木雕和砖雕居多，尤以分割内院与入口天井的垂花门为重中之重。

Built during the reign of Emperor Wanli of the Ming Dynasty, Nan residence used to be the residence of Fu Lai-jin, Deputy Surveillance Commissioner of the Ministry of Supervision of Shanxi in the Ming Dynasty. Nan residence is constructed on north-south vertical axis and from east to west three parallel courtyards feature clear functions, the main courtyard, the sub-courtyard and the ancillary courtyard. The main courtyard is for the family sacrifice and living; the sub-courtyard is for study, leisure and entertainment; and the subsidiary courtyard is the storage room and servants' room.

The gate of the residence stands in the middle of the courtyard flat, in three-door breadth. Behind the gate is a small patio, the front side of which is designed with a screen wall. Two flower-hung doors are set in the eastern and western sides of the patio respectively, serving as the entrance to the main courtyard and the miscellaneous courtyard.

In the main courtyard, its first yard — Guifu yard has a main hall in five-room breadth. The room in the middle and two rooms on its two sides are guest rooms. Rooms at the two ends are separated with a wall for the masters or elders to live. In the second yard — the Huaiyin yard of the main courtyard, both the main room and the two wing-rooms on the east and west sides are of three-room breadth. The main room is used for the worship of ancestors and the wing rooms for the young ladyship to live. In the sub-courtyard, its first yard is a private school, including the main hall, the house facing north, the door and the family chamber for worshipping Buddha. All have a width of three rooms. Through the private school yard there comes the chess yard, the main room of which is of three-room breadth. Separated from the main courtyard and sub-courtyard, the subsidiary courtyard is accessible through a flower-hung door on the east side of the gate and there is a lane between flower-hung door and the sub-courtyard. Its first yard is the miscellaneous yard, followed by the Lingxiao yard for the servants. Nan Residence boasts a large number of wooden and brick carvings, which is particularly represented by the flower-hung door standing between the inner quarters and patio at the entrance gate.

院落 The courtyard

堂楼 The main building

民居大门 The gate of the vernacular house

院门 The gate

院落民居檐廊 The eaved-corridor of the vernacular courtyard

大门照壁 The screen wall of the gate

厅堂 The hall

甘肃临夏马宅

The Residence of the Ma Family, Linxia City, Gansu Province

马宅是一座回族上层人士的宅院，位于甘肃省临夏回族自治州首府临夏市八坊东南角的三道桥，建于20世纪30年代末，由四个院落和花园组成。

马宅南临街道，原大门现已无存。进院后，穿过花园，是一座中西合璧的砖砌二门。进门后，迎面是一座砖雕照壁，然后向左转进入前院。前院为一近似正方形的四合院，是接待宾客的场所。由前院通过穿堂小院的东北角门进入正院，是主人家眷住所。上房为三层楼房，两侧为两层的东、西耳房。经穿堂小院的西北角门可进入偏院，是亲属居住的地方。

马宅的单体建筑基本上有三类：第一种中部为凹廊，两端凸出，当地称为〞虎抱头〞式。第二种前有长廊，称〞廊檐〞式。第三种仅有挑檐而无廊，当地称为〞出檐〞式。

砖雕是临夏民居特色之一，马宅之砖雕，刀法精美，线条流畅。

The residence of the Ma Family, a house for the superior class in the Hui ethnic minority, is located in Sandao Bridge, Linxia City, the capital of Linxia Hui Autonomous Prefecture of Gansu Province. Built in the late 1930's, it is a building of four courtyards and gardens.

The residence has an access to the street on the south side and its original door doesn't exist any more. Into the courtyard through the garden, there is a brick door in both Chinese and Western styles. Right behind the door is a brick-carved wall, which leads into the front yard from the left side. The front yard is a square courtyard for guests' reception. From the front yard through the door in the northeast corner of the hallway yard, there comes the main courtyard, where the owner's family live. The principal room is a three-story building and on its east and west sides are two-story penthouses. Through the door in the northwest corner of the hallway yard, one enters the wing-yard for relatives to live in.

Singular buildings here are basically categorized into three types: the first type has a concave veranda in the central part and two protruding ends, locally known as "the tiger holding head"; the second type has a front gallery, called "the eaves of veranda"; the third type only has a cornice without corridor, locally known as "eaves".

As one of the characteristics of Linxia vernacular, brick carvings in the residence show exquisite knife-carving and flowing lines.

马宅出檐式厢房 The eave pattern wing-block of the residence of the Ma family

宅中的穿堂小院 The passage way of the small courtyard in the residence

中西合璧的二门 The second gate with the oriental and western styles

墙身砖雕 The brick carving on the wall

内院入口 The entrance of the interior yard

虎抱头式主房 The main building with the tiger holding head style

宅居檐廊 The eave corridor of the mansion

宅居院落 The vernacular courtyards

照壁砖雕 The brick carving of the screen wall

入口门洞雕饰
The carving decorations of the door made of stone at the entrance

天井民居

Small courtyard houses

南方汉族民居与北方一样，平面布局都是院落式。但是，南方地区由于气候湿热多雨，多丘陵和河流，加上人口稠密，土地资源紧张，因而，民宅占地少，布局也较紧凑，房屋毗连，称为天井式民居。

南方民居有下列特征：

一是小天井。南方天气炎热，民居要求有阳光，但又怕日晒太多，特别是西晒阳光太猛，小天井可满足这要求。此外，天井还是通风、换气、采光、排水的场所，又是民宅内交通会合的地方，是传统民居中不可缺少的元素。

二是多巷道。民居建筑群中的联系主要靠通道，称为巷道。巷道有几种，露天的称巷，有的称弄，有盖顶者称廊，可见到天空又能避雨的称檐廊或敞廊，见不到天空的称内廊。廊巷不仅是民宅室内房屋之间的联系，在宅与宅之间也有巷道，称为火巷，或称夹弄，它还兼有防火通风之用。

三是敞厅暗房，这是传统民居布局中的基本元素。厅堂是民宅中最重要的和必不可少的公共活动场所，是家族文化的核心和表现，一般都位于中心部位，面向天井，厅前的格扇采用活动敞开式。家庭有重大事情时，则把格扇拆下，使厅与天井合一，成为大空间，可以容纳更多的族人。

民居建筑组合时，可纵向，可横向，也可纵横结合，类型丰富，组合灵活。天井庭院还进行绿化，外墙封闭，宅内开敞，室内外空间紧密结合。

南方民居一般都选择良好的朝向，以便获得更好的迎风条件。同时，它通过敞厅、廊巷和天井组成一个良好的通风系统。实践证明，这是南方天井式民居持久延续的主要原因之一。

天井式民居的艺术表现主要靠建筑群体艺术、空间处理、材料质感以及室内外装饰装修和家具陈设布置。

The layout of the Han people's houses in South China is basically the same as that of the courtyard houses in North China. However, since the southern region is damp and hot with plenty of rainfall, hilly land and rivers, dense population and tight land resources, a house site is generally called a small courtyard, which takes up little land, and its layout is compact, with buildings adjacent with one another.

The vernacular houses in South China have the following features:

First, the courtyards are small. It is burning hot in the south, the requirements for the houses to receive adequate sunshine and to decrease the western exposure of the hot summer sun can be satisfied with small courtyards. In addition, the small courtyard, as the indispensible element of a homestead, is the place for ventilation, change of air, natural lighting and drainage, and serves as the rally place of the transportation within it.

Second, there are many lanes and alleys. The main passages connecting the clusters of the vernacular architectures are called lanes or alleys. There are several kinds of lanes and alleys: those in the open are called lanes, and some of such lanes may be called alleyways while those with caping are called verandas. Those in which you can catch sight of the sky and take shelter from the rain and the heat of the sun are called eaves galleries or open corridors while those in which you can not see the sky are called middle corridors. There are not only lanes and alleys that function as the connection among the buildings within a homestead, but also lanes and alleys that are called fire lanes, or passageways to connect homesteads with the function of fire prevention and ventilation.

Third, open halls and concealed chambers are the basic elements in the layout of the traditional vernacular architecture. As the hall in a home is the most important and indispensible public place as the core and representation of a kin culture, it is generally located in the central part, facing the small courtyard, and the partition board in front of the hall adopts the movable and open form. When something important occurs in the family, the partition board will be removed so that the hall and the small courtyard can be combined as one bigger space to accommodate more clansmen.

Grouped flexibly with plentiful types, the vernacular architecture can be formed longitudinally or horizontally or in a crisscross pattern. The small courtyards are also afforested, the exterior wall is enclosed while the interior of the homestead is spatious so that the spaces of the exterior and the interior of the house can be tightly connected with each other.

The vernacular architecture in South China generally has very good orientations so as to obtain a better condition of facing the wind. At the same time, a very good ventilation system is formed through the open hall, the gallery or corridor and the small courtyard. Practices have proved that this is the main reason for the small-courtyard-type houses to continue in the south.

The artistic manifestation of the small-courtyard houses mainly depends on the art of architectural complex, the treatment to the space, the impression about the quality of the material and the furnishing of the furniture as well as the decoration and furnish.

江苏苏州
东山雕花大楼
Dongshan Engraved Building, Suzhou, Jiangsu Province

苏州东山雕花大楼建成于1924年，因建筑主楼的梁桁、门窗及门楼雕刻精美华丽，俗称雕花大楼。整个建筑集砖雕、木雕、石雕、铸雕、彩绘、泥塑诸艺于一体，巧夺天工，精妙绝伦，为我国雕刻艺术中的一件精品。

雕花大楼坐西朝东，平面呈多边形四合院式，两进五开间，内有门楼、主楼、花园等。大门外建有一道曲尺形的照壁，上嵌砖刻“鸿禧”，雕琢精致。

入口门楼正面为单坡瓦顶花岗石门，水磨青砖拼贴镶嵌。门楼的上、中、下枋均饰有画像砖雕，其浮雕、圆雕、透雕和谐相融。前楼大厅是大楼雕刻最多的地方，其木雕雕饰丰富，技艺精湛。

整座雕花大楼充分展现出了江南传统民间雕刻艺术的精华，它集建筑、雕刻、书画、文学、工艺、园艺于一体，因而被誉为“江南第一楼”。

Dongshan Engraved Building of Suzhou was built in 1924. It is called the Engraved Building because the beam purlins, doors and windows and the gate of the main building are engraved beautifully and exquisitely. The architecture is really a piece of cream in the Chinese engraving art since it merges such crafts as brick carving, wood carving, color decoration and clay sculpture into an organic whole as a marvelous creation excelling nature, exquisite without compare.

The Engraved Building faces east, and its plane is of multilateral four-in-one courtyard, with two rows of the width of five rooms, in which there is a gateway, the main building and the garden. In front of the entrance gate, there is a zigzag screen wall on which an exquisite brick carving with two Chinese characters “Hong Xi (meaning grand auspiciousness)” is inlayed.

The frontage of the arch-gate is a single slope roof made of tilts and a granite gate inlayed with the paste-up of levigated blue bricks. The upper, middle and lower frames of the arch-gate are all engraved with brick reliefs whose brick-carving in relief, circular engravure and fretwork are harmonious and fused mutually. The main hall of the front building has the most carvings, whose wood carving decorations are rich with exquisite crafts.

The whole building fully reflects the essence of the folk traditional carving art south of the Yangtze River, which fuses architecture, engraving, painting and calligraphy, literature, craft and gardening into an organic whole, and so it is praised as The First Building South of the Yangtze River.

门楼砖雕 The brick carving of the gate-building

雕花大楼二楼檐廊
The eave veranda on the second floor of the engraved building

宅居花园 The garden of the residence

厅堂格扇 The partition board of the hall

内院入口门楼 The gate-building at the entrance of the interior yard

浙江东阳卢宅

The Lu's Former Residence, Dongyang City, Zhejiang Province

卢宅位于东阳市吴宁镇东门外，这里丘陵起伏、河道纵横、面山环水、环境幽雅。

卢宅创建于明永乐年间，现存建筑乃明清两代相继兴建。卢宅是一座由数条南北向纵轴线组成的建筑群，房屋数千间，占地15万平方米，主要建筑是街北的肃雍堂建筑群。

进入卢宅大夫第建筑门楼，只见双扇黑漆大门，双层砖挑屋檐，旁为灰白墙面衬托，形象朴实。

肃雍堂是卢姓大族的公共厅堂，明代所建，其规模和地位十分突出。肃雍堂面阔三间，檐高脊高，廊庑式建筑。大厅与后堂用穿堂相连，形成工字形平面。肃雍堂共九进院落，是卢宅保存最完整的一条轴线。屋内梁架、柱枋、挑檐等木雕装饰工艺精湛，题材丰富，构图和谐，显示了浓郁的地方特色。

The Lu's Former Residence is located outside the east gate of Wuning Town of Dongyang City, which is diversified by hills and mountains, and a crisscross network of rivers and streams. Facing hills and mountains and surrounded with water, the residence has a very elegant environment.

The Lu's Former Residence was set up during Emperor Yongle's reign of the Ming Dynasty (1403-1424) and the existing houses were built in succession during the Ming and the Qing Dynasties. The Lu's Former Residence is an architectural complex consisting of several north-to-south vertical axes, with several thousand of houses on an area of 150,000 square meters. The main architecture is the Suyongtang complex at the north end of the street.

Entering the arch gateway of the former residence of Lu (a high official in Feudal China), you can see the double-leaf black-lacquered gate and the double-layer brick overhanging eaves of the houses against the grayish wall, with a very simple and plane appearance.

The Suyongtang Hall is the public hall for the big clan of the Lu's, built in the Ming Dynasty, with a very prominent scale and social status. It is as wide as three bays, with high eaves and a high ridge. The main hall is connected with the back hall by a hallway, form a 工-shape plane. The Suyongtang Complex consists of nine yards as the most completely preserved axis in the Lu's Former Residence. The craft of such wood-carving decorations as the beam mounts, the pillar columns, the cornices, etc. is exquisite with rich themes and harmonious designs, reflecting the strong characteristics of the local culture.

世德堂入口门楼 The gate building of the entrance of the Shide Hall

厅堂天井 The small coutyard of the hall

后院厅堂 The hall of the back yard

内院门楼 The gate-building of the interior yard

门楼雕饰 The carving decorations of the gate-building

格扇木雕 The wood carvings of the partition board

肃雍堂 The Suyong Hall

牌坊石雕
The stone carvings of the memorial gateways

出檐木雕
The wood carvings on the protruded eave

梁架木雕
The wooden beam carvings

肃雍堂后的穿堂　The passage way behind the Suyong Hall

外檐木雕 The wood carving of the exterior eave

厅堂梁架 The wooden beam of the hall

树德堂院落山墙　The gable of the Shude courtyard

大夫第大门门楼
The gate-building of the arch gateway of the ancestral hall

入口牌坊群
The complex of the memorial gateways at the entrance

浙江兰溪诸葛村

Zhuge Village, Lanxi City, Zhejiang Province

诸葛村位于浙江兰溪市，古称“高隆”，汉代诸葛亮后裔在此定居后，即按照先祖创造的九宫八卦阵式规划和建造村庄。

诸葛村自然环境好，村后高山，村前平原，可耕可樵，可渔可猎，且交通方便，适宜人居。村落位于8座小山的环抱中，其布局以钟池为中心，房屋呈向心状，向外延伸的8条巷道将全村分为8块，从而形成了内八卦布局模式。村中心的钟池为八卦式的大池塘，形似太极阴阳图形，有水的一半为阴，另一半干地为阳，池水给诸葛村增添了神秘的色彩和活泼之灵气。

诸葛村现存有明、清宅居建筑200多座，布局奇巧、结构精致。丞相祠堂建于明万历年间，建筑中庭选用了4根直径约为50厘米的松木、柏木、桐木和椿木，寓意“松柏同春”。整个祠堂翼角高翘、雕刻精美、造型庄重、气势森严。

Zhuge Village is situated in Lanxi City, Zhejiang Province and once named Gaolong in the ancient times. After the descendants of Zhuge Liang (a statesman and strategist during the period of the Three-kingdoms) settled there, the village was planned and constructed according to the Nine-Palace-Eight-Diagram battle formation created by their ancestor Zhuge Liang.

Zhuge Village boasts itself for a very good natural environment with high hills behind and plains in front of it where people can make a living either by farming or by gathering firewood, by fishing or by hunting. Furthermore, it has a very convenient transportation so that it is appropriate for inhabitancy. The village is surrounded with 8 hills and its layout takes the Zhong Pond as the center and the houses are of the radial form. The 8 lanes spreading outward divide the whole village into 8 sections so as to form the pattern of interior Eight-Diagram battle formation. The Zhong Pond in the center of the village, a big pond of Eight-Diagram battle formation, looks like a diagram of supreme pole with Yin and Yang, the half with water as Yin and the other dry half as Yang. The water in the pond adds mystical coloration and vivid anima to the village.

More than 200 vernacular houses built during the Ming and Qing Dynasties exist in Zhuge Village, whose layouts are ingenious and structures exquisite. The Ancestral Temple of the Prime Minister (in ancient China) was built during the reign of Emperor Wanli (1573-1620) of the Ming Dynasty. Four logs with the diameter of 50 centimeters (a pine, a cypress, a tung and a Chinese toon) were selected for building the Middle Hall, with the implied meaning of Song Bai Tong Chun (meaning “may you live long and remain strong like the evergreen pine and cypress”). The wing corners are all raised with exquisite carvings, solemn looks and imposing momentum.

宅第入口 The entrance of the homestead

诸葛村上塘建筑群
The architectural complex of the upper pond in the Zhuge Village

雍睦堂 The Yongmu Hall

林杏小院门洞 The gateway of the Linxing Small Courtyard

进士第门楼 The gate-building of the scholar's homestead

高低错落的马头墙 The horse-head gable with different heights

林杏小院花园 The Garden of the Linxing Small Courtyard

浙江永嘉
岩头丽水街

Lishui Street, Yantou, Yongjia Country, Zhejiang Province

岩头丽水街位于楠溪江中游西侧，始建于唐初。塔河庙由庙、河、戏台、接官亭、古树等组成，是村民的大型户外活动中心，堪称风景园林。

东园的蓄水堤上，建有一条古商业街——丽水街。300多米长，90多座相连的一二层店屋，对水一字排开，临水有美人靠，历史上为盐商必经之路，兼有小憩和商业作用，也是一处极具特色的景观街道。

街南端是寨墙的南门，门边高阶上有凉亭，离亭50米处有一座厚重又灵巧的花亭（又名接官亭），四方二层，底层木板墙，二层外挑，通体花窗，重檐攒尖顶，朴素庄重，与丽水街相互呼应。

The Lishui Street of Yantou, located on the west bank of the middle reaches of Nanxi River, was built in the early Tang Dynasty. The Tahe Temple on it, being composed of the temple itself, the river, the stage, the official-reception pavilion, the age-old trees etc., is a large-scale outdoor activity center for the local villagers. It deserves the name of a landscape garden.

On the reservoir embankment in the East Park, there built an ancient commercial street — the Lishui Street. It is more than 300 meters long, with 90 multi-faceted one-story or two-story shops, lined up across the river. Near the water is the meiren kao (a chair wins its name for being seated by beauties and leaned along their waist). In the history, the place was the only passage for salt merchants, who could have a rest or do business here. It is also a very unique street with a pleasant landscape.

At the south end of the street is the South Gate and there is a pavilion on the terrace nearby. Fifty meters away from the pavilion is a sound and delicate flower pavilion (also known as the official-reception pavilion), four-sided and two-story, wooden wall on the first floor and out-stretching eaves on the second floor. With transparent flower windows and heavy eaves, the pavilion features simplicity and solemn elegance, matching well with the Lishui Street.

接官亭 The official reception pavilion

丽水街沿河长廊 The gallery of the Lishui Street along the river

长廊美人靠 The meiren kao gallery

商业古街 The historic business street

浙江永嘉苍坡村

Cangpo Village, Yongjia County, Zhejiang Province

村巷呈八卦形，以方形环状的鼓盘巷为中心，向四方布置八条路，通向村外，并渗进文房四宝理念。水池为砚，一条直街正对村外笔架峰为笔，池旁置长条石为墨，全村用地为纸。

村落完成于南宋，苍坡溪门（当地称库门）为牌楼木构，六柱三架二步梁。屋面重檐悬山顶，明间屋面高出，形成三山式，小青瓦屋面，檐口置勾头滴水，花脊两端饰龙凤吻兽及垂兽，明间两金柱，次间各设两角柱，柱头科为十一踩四抄双下昂，攀间额枋设四组平身科，为九踩四抄双下昂式。

东西池、仁济庙大宗祠组成村民活动中心，活动中心东南角立望兄亭，遥对南边方岙村的"送弟阁"，流传着李氏七世祖李嘉木、李秋山兄弟的一段佳话。

池东北端的"水月堂"为一组准园林建筑，始建于宋徽宗宣和年间（1119—1125年）。苍坡村的规划和建筑都流溢着宋元风格。

Lanes in the village are arranged in the shape of Eight Diagrams. Centering on Gupan Lane in a square-ring shape, eight roads extend in four directions, infiltrated with the idea of four treasures in the study. The pool symbolizes a Yan (ink-slab); a straight road facing the Bijia (penholder) peak outside the village is the pen; a long stone by the side of the pool is the ink, and the land in the village is the paper.

The village was completed in the Southern Song Dynasty. The Cangpoxi Gate (locally called the framed door) is of wooden arch, six columns, three frames and two beams. With overhang roof eaves, the facade of the room in the middle is higher than that of others, forming a "山"-shape. A small tile roof is set with hook along the mouth for water dripping. Both ends of the ridge are decorated with dragons, phoenixes, flowers and other animals. Two golden columns stand in the room in the middle and in the room next to it there are two corner columns. Tou-Kung at the top of each column is with 11 Cais — 4 Chaos and 2 Angs. 4 intermediate sets are 9 Cais — 4 Chaos and 2 Angs.

The east and west pools and the Renji ancestral temple serve as the activity center for the local villagers. At the southeast corner of the activity center stands a pavilion named "Wang Xiong Ting (waiting for elder brother)", which pairs with the pavilion of "Song Di Ge (seeing off younger brother)" in Fang'ao Village in the south, which tells the story of brotherhood between Li Jia-mu and Li Qiu-shan, the ancestors of the seventh generation of the Li clan.

At the northeastern end of the pool, the "Shui Yue Tang (water-moon hall)" as a quasi-garden building, was built in the reign of Emperor Huizong in the Song Dynasty (1119-1125). The planning and construction of the Cangpo Village reflect the architectural style in the Song and Yuan dynasties.

村口古道 The ancient road at the mouth of the village

村落水景 The water scene of the village

砚池建筑群 The complex of architecture of the ink-slab pool

临池美人靠

The meiren kao close to the pond

浙江永嘉芙蓉村

Furong Village, Yongjia, Zhejiang Province

楠溪江中游岩头镇的芙蓉村，因位于芙蓉山脚下而得名。始祖为河南开封人，唐朝“安史之乱”时迁徙至瑞安长桥、福建长溪，九世祖从长溪迁至此。宋代实行耕读政策，楠溪江流域为理想的耕读之地，芙蓉村学风鼎盛，陈氏成为了“簪缨鹊起，甲第蝉联”的名门望族，

村落曾毁于战火，元至正年间重建，应用“以象制器”的方法，“七星八斗”的布局模式（斗为水塘、水池，星为道路交叉口上的石台），全村方砚形，用粗石干砌寨墙围合。现存明清民居50余处，以司马第、书院、将军屋最为有名。

Located in Yantou Town in the middle reaches of the Nanxi River, Furong Village is named after its location at the foot of the Furong Mountain. The ancestors of the village came from Kaifeng of Henan Province, and relocated in Rui'an, Changxi of Fujian Province after the An-Shi Rebellion in the Tang Dynasty. The ancestors of the ninth generation migrated from Changxi and finally inhabited here. Thanks to the policy of farming and study advocated in the Song Dynasty, the Nanxi River area became an ideal place for farming and study and Furong Village enjoyed a fine tradition of study. As a result, the Chen clan became a renowned family for "well-known scholars and high-rank officials".

Destroyed in flames of war, the village was rebuilt in the first year of the Yuan Dynasty, adopting the Chinese ancient theory of "function is embodied by a form" and "seven-star and eight-dou" in its arrangement, the pond and pool standing for the dous and the stone desks at road intersection for the stars. The entire village is in a square shape and enclosed by walls of rubbles. More than 50 vernacular houses from the Ming and Qing dynasties are still standing, of which the General's Residence, the school and the General's House are most famous.

民居入口 The entrance of the vernacular houses

村落中心的芙蓉池、芙蓉亭 The Furong Pond and the Furong Pavilion at the center of the village

芙蓉书院 The Furong School

书院大门 The gate building of the school

浙江平阳顺溪民居

Shunxi Vernacular Houses, Pingyang County, Zhejiang Province

平阳顺溪民居位于溪畔，溪流两岸群山耸立，层峦叠嶂，树木青翠，现保存基本完整的古民居有10幢，其中8幢为陈氏（育球）七个儿子于清乾隆年间相继建造而成。

建筑总特点为轻盈、透薄、大屋顶、大挑梁，多为横向发展的四合院组成，用回廊连接，以立面为特色，有的门前置有旗杆石，建筑多有门额，开间多，明间阔。

主政第三进九间，横三院式，主立面横长，门前有旗杆石，主院二楼设走马楼，做拼花美人靠，窗花丰富。户侯第为一品字形院落，前设台门，台基用断砌造，门外两侧各立两对旗杆石，正厅、厢房皆为两层楼，上层檐柱退进设走马廊，漏墙和瓦当极富特色。

另外，位于镇郊的陈有相大屋，二进九间合院式，木构建筑，由砖门台、门厅、正厅、两厢及附屋组成，梁、枋、雀替、斗栱上无处不雕，雕刻工艺十分精湛。

Shunxi vernacular houses are located along the stream banks, and on the two sides of the stream are towering mountains, range upon range covered with exuberant vegetation. Today, ten buildings are well preserved, of which eight were constructed during the reign of Emperor Qianlong of the Qing Dynasty by the seven sons of Chen Yu-qiu.

The construction is light and sheer, boasting large roof and large cantilever beam. It is composed of courtyards in east-west orientation linked by corridors. Featuring on the facade, some courtyards have flag stones standing in the front gates and doors in multi-room breadth.

The main courtyard has three halls and nine rooms, consisting of three yards in east-west orientation with a flag stone in front of the main façade. Its second floor is equipped with corridors, meiren kao (banisters that win the name for beauties to lean along their waist) and grille windows with rich pictures. The residence of the Marquis is a 品-shaped courtyard. Its gate is designed as a platform door and its foundation adopts section construction. On both sides of the gate stand two pairs of flag stones. The main hall and wing-hall are of two-story buildings, with eave columns on the upper floor setting back to create a Zouma (gallop or trot along on horse back) corridor. Its leaking walls and roof tiles have rich characteristics.

Besides, the big house of the Chen Youxiang outside the town is a courtyard with two halls and nine rooms, constructed of wood. It is composed of the brick gate, the door hall, the main hall, two wing-halls and a subsidiary house. Its beams, tiebeams, sparrow braces and brackets are all covered with carvings of superb craftsmanship.

偏院 The wing yard

陈氏主政第二楼檐廊 The eave veranda on the second floor of the main building of the Chen Clan

内院 The interior yard

花厅 The flower hall

侧院 The side yard

浙江平阳青街民居

Qingjie Vernacular Houses, Pingyang County, Zhejiang Province

平阳青街为畲族乡，三条溪水在此汇合。民居多为合院。如建于明万历二十五年的水尾池氏老宅，由三个横向发展的四合院组成，主立面有十一间，尺度近人，装饰简朴，为温州典型的多院落式长屋。

李氏大屋位于老街南端，清代建筑，为面厅式四合院，正屋九间，两厢各三间，两层，正厅很阔（约6–7米），院落很大，约18米×18米，环院落一圈为连廊（底层）和走马廊（二楼）。沿街的外墙用粗石砌筑且不开窗，该屋大小木装饰（修）简单使用，其挑檐、铺作极具特色。

Qingjie is the home of the She ethnic minority, and three streams converge here. Most vernacular houses are in the courtyard-type. For example, the residence of the Chi clan built in the 25th year of Emperor Wanli in the Ming Dynasty is composed of three courtyards in east-west orientation. The main façade has eleven rooms with reasonable width and simple decoration. It is the typical multi-courtyard long-house in Wenzhou.

The big house of the Li clan, located in the south end of the street, is a hall-style courtyard of the Qing Dynasty. Its main hall has nine rooms while two wing-halls have three rooms. The two-story building has a wide main hall about 6-7 meters wide and a large courtyard of about 18 meters in both width and length. Surrounding the compound is the veranda on the first floor and Zouma corridor on the second floor. Walls along the street are built of rubbles without any window. The house is decorated with wood carvings in various sizes, simple and easy to use, and its brackets are of unique features.

李氏司马第入口檐廊
The eave veranda at the entrance of the Residence of Minister of the Li Clan

内院 The interior yard

3

门厅 The door hall

安徽黟县宏村

Hong Village, Yixian County, Anhui Province

宏村位于黟县城东北11公里处，坐北朝南，背倚雷岗山，溪河和羊栈河从村旁穿过。全村以半月形的月塘为中心，仿牛的形状进行村落布局，别出心裁，独具匠心。村西头的雷岗是牛首，村口的两棵参天古树是牛角、前后四座横跨吉阳河的桥梁谓牛腿，数百幢明清古建筑为牛身，环绕全村盘曲流过家家户户的水渠曰牛肠，村中半月形池塘是牛胃，村南的南湖是牛肚，整座牛形村充分体现了宏村人的聪明才智。

全村现有明、清民居132幢，其中以承志堂最为典型。承志堂建于咸丰五年（1853年），为三进两层三开间的木结构建筑，中轴线上建有前院、福堂、寿堂三组主要建筑，依次递进。东部为花园，西部观鱼厅，水院与整村的水系连通。承志堂的木雕、石雕、砖雕十分精美，有“雕楼”之称。

宏村的村外山清水秀，红杨翠柳，而村内则清渠绕户，终年清澈。

Hong Village is located 11 kilometers northeast of Yixian County, facing south, lying against the Leigang Mountain and the Xi River and the Yangzhan River flowing by the village. Taking the half-moon-shaped pond in the village as the center, the village is arranged in imitation of a cattle shape, full of ingenuity and originality. Leigang at the western end of the village is the cattle head, two towering trees at the mouth of the village being the cattle horns, four bridges across the Jiyang water being the cattle legs, hundreds of ancient buildings of the Ming and Qing dynasties being the cattle body, coiled aqueduct flowing through each household being the cattle intestines, the half-moon-shaped pond being the cow stomach, and the lake in the south of the village being the cattle tripe, fully reflecting the intelligence of the Hong villagers.

There are 132 existing vernacular houses of the Ming and Qing dynasties in the village. Of them, the Chengzhi Hall is the most typical one. The hall, built in the 5th year of Emperor Xianfeng of the Qing Dynasty (1853), is a three-hall and two-story wooden structure in three-room breadth. On its central axis three principal buildings: the front yard, the Fu (blessing) Hall, and the Shou (longevity) Hall are deployed. The east is a garden and the west is the Guanyu (fishing pool) Hall. The water garden is connected with the water system of the entire village. Because of the extraordinary wood, stone and brick carvings there, the Chengzhi Hall is also reputed as a carving building.

Outside the village, green mountains, luxuriant trees and colorful flowers create a wonderful landscape and inside the village, a clean canal runs through each household, singing all year long.

村中心半月形水池 The half-moon-shaped pond at the village center

宅居前院 The front yard of the residence

古村落街巷 The streets and lanes in the historic village

巷道　The lanes

八字形大门 The 八 -shaped gate

小院水榭 The waterside pavilion of the small courtyard

承志堂正厅 The main hall of the Chengzhi Hall

宅居内院
The interior yard of the residence

檐廊雕饰
The carving decorations of the eave veranda

牌楼式大门
The pailou-shaped gate

安徽黟县
南屏村
Nanping Village, Yixian County, Anhui Province

南屏村位于黟县县城西南4公里处，因村西南背倚南屏山而得名。

南屏村高墙深巷，全村1000多人，却有36眼井，72条巷，300多幢明清古民居。全村巷弄，长短不一，弄弄相通，拐弯抹角，纵横交错，有江南迷宫之称。

南屏村至今乃保存有相当规模的宗祠、支祠和家祠，被誉为中国古祠堂建筑博物馆。叶氏宗祠〝叙秩堂〞位于南屏村中心，始建于明成化年间，坐东朝西。大门两侧有一对一人多高、用黟县青石精雕细刻的巨大石鼓，乌亮厚实，非常威严。从大门口可以一直望见上厅享堂的香火案壁，显得轩敞又幽深，使人顿生肃穆、神秘之感。

位于村庄上首的半春园，建于清光绪年间，是村中一所私塾庭院。园中种植了大量名贵花木，尤以梅花为最，故别名为梅园。

Nanping Village, located four kilometers southwest of Yixian County, is named after the mountain it lies against in the southwest.

Boasting high walls and deep lanes, Nanping Village has more than 1000 villagers and there are 36 wells, 72 lanes and more than 300 vernacular houses of the Ming and Qing dynasties. Lanes in different lengths interlinked with each other and running in all directions are found everywhere in the village, winning it the name of labyrinth in the south Yangtze.

Even today, a considerable scale of temples, shrines and ancestral halls still stand in Nanping Village which is known as the architectural museum of Chinese ancient shrine. The Ancestral Hall of the Ye clan "Xuzhi Hall", located in the center of Nanping Village and facing west, was built during the reign of Emperor Chenghua of the Ming Dynasty. A pair of huge stone drums carved out of the bluestone of Yixian County about a person's height stand on both sides of the gate, glossy and dignified. From the entrance hall, the incense case in the main hall can be beheld and the whole building is spacious, bright and deep, creating a sudden and devastating silence and mystery.

The Banchun Garden at the entrance of the village, built during the reign of Emperor Guangxu of the Qing Dynasty, is a private school courtyard. There plant a large number of rare flowers and trees, especially plum for the most, thus it is also named the Plum Garden.

叶氏支祠　The shrines of the Ye Clan

奎光堂 The Kuiguang Hall

民居建筑群 The complex of the vernacular houses

街巷 The lanes

民居大门 The gate of the vernacular houses

村落巷道 The lanes of the village

大门 The gate

上叶街 The Shangye Street

天井小院 The small courtyard

敬贤堂 The Jingxian Hall

民居前院 The front yard of the vernacular houses

安徽黟县关麓村

Guanlu Village, Yixian County, Anhui Province

关麓村位于安徽黟县西南武亭山麓，是黟县聚族而居的大村，距今已有一千多年的历史。

清朝中叶，汪氏经商积财，回乡大兴土木，建筑楼房连片，飞檐翘角、画栋雕梁、造型十分精美。

关麓村有一片古建筑群，保存较好的明清民居有100多幢，其中最引人注目的8处为村核心建筑，被称为“关麓八大家”。建筑群的特点是：八家宅第门户相通，楼与楼相连，成为一个整体，但八户又各自独立，自成一家，互不干涉。在建筑形式上，风格统一，有皖南地方特色。

本村民居大门为当地传统的门楼形式，青石门框，内为双扇黑漆大门，门楣做成出挑式，短檐单坡顶，它与粉墙组合，黑白分明，形象突出。

Guanlu Village, a big village where some clans live, is located at the foot of the Wuting Mountain on the southwest of Yixian County, Anhui Province, with a history of more than 1000 years.

In the middle period of the Qing Dynasty, when the Wang's accumulated much wealth by engaging in trade, they returned to their hometown and started architectural work on a large scale. They built a large number of houses with upturned eaves and bent corners, carved beams and painted rafters and their models are perfectly elegant.

Guanlu Village has a complex of ancient architecture and more than 100 houses built during the Ming and Qing Dynasties are preserved relatively well, among which the core architecture are the eight most eye-catching ones named the Eight Guanlu Residences. The feature of the complex is that the eight residences are connected and lead to one another as a whole though they are independent of each other without mutual interference. The integrate style in the form of the architecture is an obvious local character in the south Anhui.

The entrance of the village is a locally traditional gateway with blue stone as its frame and two-leaf black doors. The lintel of the gate is overhanging, with a short-eave single slope roofing, combined with the white-washed wall, to form an outstanding image of clear distinction between black and white.

关麓村民居 The vernacular houses of the Guanlu Village

民居大门 The gate of the vernacular houses

安徽歙县
棠樾村
The Vernacular Houses of Tangyue Village in Shexian County, Anhui Province

棠樾古村位于歙县，自宋代以来已有800余年历史。

在歙县，最引人注目的是棠樾古村的牌坊群，七座牌坊逶迤成群，古朴典雅，蔚为壮观。牌坊中，明代三座，清代四座，用石料建成，以质地优良的"歙县青"石料为主，高大挺拔，气宇轩昂，既不用钉又不用铆，以石与石之间的巧妙结合历千百年而不倒。连续的牌坊相互唱和，以石头的群体记录了地域文化特色。

棠樾民居基本上是高大的墙垣围成长方形的天井式民居，两层楼房。较大的宅居在侧边布置庭园，园内植花草树木，有漏窗花墙相隔。民居造型丰富，双坡顶、白粉墙、上覆青瓦、清雅朴素，两山跌落式马头墙为其主要特征。

The historic Tangyue Village is situated in Shexian, with a history of more than 800 years since the Song Dynasty.

In Shexian County, the most eye-catching building is the complex of the memorial gateways of the historic Tangyue Village. Seven memorial gateways are winding into a complex, simple, unsophisticated and elegant, affording a magnificent view. Among the memorial gateways, mainly made of Shexian Blue stone, three were built during the Ming Dynasty and four during the Qing Dynasty. All of them are huge, tall and straight, with imposing appearances. Without nails or rivets, the stone blocks are integrated with one another skillfully so that the complex of memorial gateways has been standing there for around a thousand years. The successive memorial gateways are in perfect harmony, having recorded the cultural characteristics of the region with the group of stone blocks.

The vernacular architecture of Tangyue Village are mainly rectangular small-courtyard two-storied houses surrounded with tall walls. The relatively residences have gardens on their sides separated by tracery walls, in which flowers and plants are grown. The houses have abundant models, with double-slope roofs and white walls covered with blue tiles, elegant and simple. The piled horse-head walls on both gables are the main features of the houses there.

鲍宅厅堂外檐
The outer-eaves of the Bao-Clan Hall

敦本堂祠 The temple of the Dunben Hall

棠樾牌坊群 The complex of the Tangyue memorial gateways

祠堂厅堂 The hall of the ancestral temple

安徽歙县
唐模村
Tangmo Village, Shexian County, Anhui Province

唐模古村位于安徽黄山风景区的南麓，处于连绵起伏的群山环抱之中。它在规划布局整体村落及营造方面，由于历史、经济、自然条件的特殊性，得以保存了众多品位很高的徽派古建筑。

唐模古村特色体现为两大部分：

其一为水街。水街长达1100余米，两岸分布着近百幢徽派民居，并形成了夹道而建的街道市井。沿街有40余米长的避雨长廊，廊下临河设有美人靠，供人来往休息。街区周围有古桥、古祠、古村、古井，可谓古韵悠悠。

其二为水口园林——檩干园，它始建于清初，占地十余亩，是儒家文化的集中体现。园内三塘相连，有三潭映月、湖心亭、笠亭等胜景。更令人流连忘返的是园中镜亭内的宋、元、明、清18位名家的真迹石刻，镌刻精致，气势恢宏。

The historic village of Tangmo, located in the southern foot of the Huangshan Mountain in Anhui Province, is surrounded by waving mountains. Due to the historic, economic and natural conditions, the overall layout, planning and construction of the village has helped it to preserve many high-quality ancient buildings of the Hui School.

The characteristics of the ancient village are embodied in two parts:

One is the Water Street. Up to 1100 meters, the Water Street is lined with hundreds of Anhui vernacular houses on two banks, forming streets and marketplaces. Along the street there is a 40-meter-long gallery to provide people sanctuary from the rain, with meiren kaos (banisters that win the name for beauties to lean along their waist) set along the river under the gallery for rest. Surrounding the streets are the bridge, the ancestral hall, the village and the cell from the ancient time, permeating an ancient rhyme.

The other is the Water Garden — the Lin'gan Garden. Founded in the early Qing Dynasty, the garden, covering more than ten mu, is the embodiment of the Confucian culture. Three ponds in the park are jointed with each other, creating breathtaking scenes of three ponds mirroring the moon, the pavilion in the lake's center and the Li pavilion. The most attractive scenes in the park are the authentic stone-engravings of the works from 18 masters in the Song, Yuan, Ming and Qing dynasties, delicate and magnificent.

唐模水街民居 The vernacular houses of the Tangmo water Street

肇敦堂 The Zhaodun Hall

檀干园内景 The interior scene of the Tanggan Garden

庭园连廊 The garden veranda

村口檀干园园林建筑
The Tangan Garden architecture at the mouth of the village

连廊 The veranda

厅堂前院 The front yard of the hall

水街 The Water Street

高阳桥 The Gaoyang Bridge

高阳桥上看水街 Looking at the Water Street from the Gaoyang Bridge

江西婺源
李坑村
Likeng Village, Wuyuan County, Jiangxi Province

“坑”在赣中就是溪的意思，婺源所有的村庄都建在碧波荡漾、纵横多姿的溪流边上。秋口镇李坑村坐落在一条狭长的山坞之中，建于北宋年间，村落群山环抱，山清水秀，风光旖旎。村内260多户大都沿河而居，每户门前都有一座青石板桥，真是名副其实的小桥流水人家。

李坑村民居属于徽派建筑，除了具有徽派建筑典型的粉墙、飞檐、翘角外，木、石、砖三雕也可称“三绝”。村内明清古建遍布，民居宅院沿苍漳依山而立，粉墙黛瓦，参差错落。村内街巷溪水贯通、九曲十弯。青石板道纵横交错，石、木、砖各种溪桥数十座，沟通两岸，构筑了一幅小桥、流水、人家的美丽画卷，是婺源古村落中的一颗灿烂明珠。

In Jiangxi Province, “Keng” means a brook. All the villages in Wuyuan are built along the transparent and zigzag brooks. Likeng Village of Qiukou Town, situated in a narrow mountainous dock, was built in the Northern Song Dynasty. It lies in the arms of green mountains, enjoying a beautiful scenery. Most households of more than 260 in the village inhabit along the river and a bluestone bridge is accessible to every house, creating a vivid picture of a small bridge over the flowing stream, and cottages.

Likeng vernacular houses belong to the Hui-school architecture. Apart from the typical features of the Hui-school architecture such as the powder wall, the eaves and cornices, the carvings of wood, stone and brick can also be called three marvelous techniques. The ancient buildings from the Ming and Qing dynasties can be found anywhere in the village, which are constructed along the Cangzhang Mountain, covered with eye-pleasing walls and tiles. They are scattered irregularly. Within the village, streets, lanes and streams traverse and cross each other. There are criss-cross roads paved with bluestone and dozens of stone, wood and brick bridges fly over the stream, creating a splendid picture-scroll of small bridge, flowing water and houses. It is a brilliant pearl in the ancient villages in Wuyuan.

李坑水街 The Water Street of the Likeng Village

石拱桥 The stone arch bridges

水埠 The wharf

河道 The stream way

大夫第 The ancestral hall

江西婺源理坑村

Likeng Village, Wuyuan County, Jiangxi Province

理坑原名理源，位于婺源县城56公里的沱川乡。村建于北宋末年，村人好读成风，崇尚朱子理学。几百年来，这偏僻山村秉承勤学苦读之风，人才辈出。

理坑府第与商家宅第不同，更讲究内外布局正统规范，显示官宅的气派，至今仍保存完好的古宅有明崇祯年间余自怡的官厅、明万历年间余懋学的尚书第、清顺治年间余维枢的司马第、清道光年间余显辉的诒裕堂、花园式的云溪别墅、园林式建筑花厅、作为我国古代民间建筑艺术珍品的经义堂等，这些府第建筑造型雍容大方，外观粉墙黛瓦，内部雕饰精湛，布局科学合理，冬暖夏凉，是建筑艺术的博览园。理坑因完整的明清古宅群和深厚的人文底蕴，成为婺源古村落中的一个典型代表。

Likeng, originally called Liyuan, is located in Tuochuan town, 56 km away from Wuyuan City. The village was built at the end of the Northern Song Dynasty when the villagers loved studying and advocated Zhu Neo-Confucianism. For centuries this remote mountainous village has upheld the tradition of study and bred a galaxy of talented people.

Different from the residences of merchants, the residences in Likeng place more importance on the formal specification of the inside and outside layout, so as to display the grandeur of the official residence. Up to now, the well-preserved ancient residences include the government office of Yu Zi-yi during the reign of Emperor Chongzhen of the Ming Dynasty, the residence of Minister Yu Mao-xue during the reign of Emperor Wanli of the Ming Dynasty, the residence of General Yu Wei-shu during the reign of Emperor Shunzhi of the Qing Dynasty, the Yi Yu Hall of Yu Xian-hui during the reign of Emperor Daoguang of the Qing Dynasty, the garden-style Yunxi villa and the flower hall, as well as the Jingyi Hall, the architectural treasure in the ancient China. These mansions boast graceful architectural style, eye-catching outer appearance and delicate internal carvings and decorations. With a scientific and rational layout, they are cool in the summer and warm in the winter, showcasing a variety of architectural art. Thanks to its completely-reserved ancient architectural complex from the Ming and Qing dynasties and the profound cultural heritage, Likeng Village has become a typical example of the ancient village in Wuyuan.

民居厅堂 The hall of a vernacular house

绣楼　The embroidery building

村落水口　The water outlet of the village

古街深宅 The mansion with many courtyards in the old street

江西婺源思溪村
Sixi Village, Wuyuan County, Jiangxi Province

思溪村位于婺源思口乡，建于南宋庆元五年（1199年），村落状似船形，因鱼思恋溪水而得名。村落背山面水，嵌于锦峰绣岭、清溪碧河的自然风光之中。进入村子须经横跨小河的风雨桥廊，为明代所建的"通济桥"。

现村中保存有明清民居30多幢。清代商家住宅有"振源堂"、"承裕堂"、"承德堂"、"孝友兼隆厅"等。建于清雍正年间的敬序堂，厅堂悬挂"敬序堂"鎏金匾额，厅内藻井为长方形，古色古香。建筑群由六个天井组成，书斋朝向庭园，简朴幽静，楼房有回廊护栏。

雕刻工艺精湛。清乾隆年间所建俞氏客馆，隔扇门上分别镌刻着由96个不同字体的"寿"字组成的"百寿图"，隔扇门上下还雕镂有人物戏文、鱼虫花鸟、水榭楼台等图案，堪称木雕精品。

Located in Sikou Town of Wuyuan County, Sixi Village was built in the 5th year of Emperor Qingyuan's reign of the Southern Song Dynasty (1199). In the shape of a boat, it is named for fish longing for streams. Lying against a mountain and facing water, the village is embedded in the embroidery natural scenery of green mountains and blue water. A bridge across the river and leading to the village is "Tongji Bridge" built in the Ming Dynasty.

More than 30 vernacular houses from the Ming and Qing dynasties are preserved in the village. Residences for merchants in the Qing Dynasty include the "Zhenyuan Hall", the "Chengyu Hall", the "Chengde Hall", the "Xiaoyou Jianlong Hall" and so on. In the Jingxu Hall which was built during the reign of Emperor Yongzheng of the Qing Dynasty there hangs a gilt plaque of the Chinese characters "Jing Xu Tang (Jingxu Hall)". The patio in the hall is in rectangular shape, full of antique flavor. The building combines six yards with patios; the study faces the garden, quiet and simple; and verandas are constructed surrounding the building as the fence.

As an example of the exquisite carving, the Guest Hall of the Yu clan, built during the reign of Emperor Qianlong in the Qing Dynasty, has a shelf fan door inscribed with the Chinese character "Shou (longevity)" in 96 different fonts, creating a "100-Shou Picture". On the upper and lower ends of the door, there are engraved pictures of characters, dramas, fish, insects, flowers, birds, water towers, pavilions and other patterns. It is no doubt a masterpiece of wood carving.

风雨桥美人靠
The meiren kao on the rain-proof bridge

村口通济桥 The Tongji Bridge at the mouth of the village

入口门楼装饰 The decoration of the gate-building at the entrance

村落巷道 The lanes in the village

院落休息亭
The resting pavilion in the courtyard

江西赣县
白鹭村
Bailu Village in Ganxian County, Jiangxi Province

白鹭村位于赣州市赣县北部，距赣州市63公里，早在商周时期就已形成人居村落，此后日渐繁荣。村落面积有0.2平方公里，沿着鹭溪呈月牙形分布。村里的四条主要街道，极似一个大的“丰”字，形成了村落景观“白鹭十景”：天一池、二义仓、三元官、四逸堂、五福第、六角亭、七姑庙、八角井、九成堂、十字街。

古色古香的青砖黑瓦建筑群多为明、清两代所建，百年以上的民居就有140多栋。宗祠民居雕梁画栋，大小天井错落有致，正厅偏厅相得益彰，住房杂屋秩序井然。

白鹭祠堂封火山墙，飞檐翘角，威武堂皇，气势非凡。恢烈公祠为白鹭最大的建筑群，前后三进，大小天井16个，与“九井十八厅”相当。王太夫人祠，是以女性名字命名的祠堂，在过去封建社会极其少有，因王太夫人乐行善施，为其立祠祭祀，王太夫人祠也是赤贫子弟的私塾。

Bailu Village is located in the north of Gan County, 63 km away from Ganzhou City. As early as in the Shang and Zhou dynasties, a village for man's habitat took shape, later developing into prosperity. The village covers an area of 0.2 square kilometers and spreads in crescent shape along the Luxi Stream. The four main streets in the village form a large Chinese character “丰(abundance)”. Its landscape forms “Bailu Shijing (ten scenes)”: one pool, two granaries, three Gods, four-leisure hall, five blessings, six-point pavilion, seven-auntie temple, octagonal well, nine-accomplishment hall and Cross Street.

Antique brick buildings covered with black tiles were mostly built in the Ming and Qing dynasties and more than 140 buildings were built over a century ago. Ancestral houses have carved beams and patios are orderly arranged. Main halls and wing-hall complement with each other and main houses and miscellaneous houses are placed in good order.

The Bailu ancestral hall is constructed with fire gables and flying rafters, looking grand and magnificent. The Huilie Ancestral Hall, as the largest architectural complex in Bailu, is a three-hall building with 16 patios in various sizes, quite on a par with “nine-patio and eighteen-hall”. Madam Wang's Temple is an ancestral hall named after a female, which was extremely rare in the feudal society. To honor the lady's generosity and hospitality, the ancestral hall was built and later also served as the private school for children from poor families.

绣楼 The embroidery building

白鹭村民居街巷
The lanes of the vernacular houses in the Bailu Village

四逸堂 The Siyi Hall

王太夫人祠外墙面
The exposed wall of the Madam Wang's Temple

兰胜堂 The Lansheng Hall

江西吉安
钓源村
Diaoyuan Village in Ji'an City, Jiangxi Province

吉安市吉州区钓源村，是北宋大文学家、政治家欧阳修的后裔、宗裔聚居地。明清之际，钓源在吉安（古称庐陵）地区享有"小南京"之美誉。

钓源村依太极、八卦图形成布局，以东西走向的"S"形长安岭为屏，村落依照太极图的鱼尾形坐落在"S"形的两个湾里。全村无论街、道、巷，无一取直。走进村庄，一条条青石板路，一段段卵石路面曲折、宽窄不一地呈现在你面前。钓源村现有上百栋明、清建筑，包括庙观、祠堂、书院、别墅、民居等。

Diaoyuan Village in Jizhou District, Ji'an City, is the settlement of the descendants of Ouyang Xiu, the greatest writer and statesman in the Northern Song Dynasty. During the Ming and Qing dynasties, Diaoyuan was reputed as the "small Nanjing" in Ji'an (ancient Luling).

Diaoyuan Village is planned in accordance with the Taiji Diagram and Eight Diagram. Taking the S-shaped Chang'an ridge running east to west as screen, the village is located in the two bays in the fish-tail-shaped Taiji Diagram. No straight street, road or alley can be found in the village. Into the village, bluestone-paved roads and gravel roads one after another take twists and turns and vary in width. Today, over a hundred buildings from the Ming and Qing dynasties have survived, including the temples, ancestral halls, schools, villas, vernacular houses and so on.

七星伴月形水塘
The pond in a shape of seven stars around the moon

钓源村中心民居建筑群
The complex of architecture at the center of the Diaoyuan Village

村前花桥 The flower bridge in front of the village

福建武夷下梅村

Xiamei Village in Wuyi City, Fujian Province

位于武夷山的下梅村是以茶叶集市发展起来的一个小村落。当溪自西向东贯穿全村，流入梅溪，再汇入崇阳溪，形成了以溪水为交通，舟楫便利的水运优势。清初，此地是武夷山茶市的经贸活动中心，外地商贩纷纷到此采购。至今，下梅人仍以农耕和制茶为主要营生。

溪旁堤边设踏步，供汲水与洗涤之用，构成以水为中心的商业街市。沿街有民宅、商肆小店、小庙、宗祠、酒店、作坊、檐廊、水井等，沿溪一侧有与溪岸平行或悬挑的休闲靠椅。

村中保留下来30余幢明清古民居建筑，有巨商的豪居，官宦的府第，儒生的学堂，最具代表性的有邹氏大夫第、邹氏家祠、程氏隐士居、西水别业、方氏参军第等建筑群。建筑砖木结构，东阁西厢、楼台歇屋、天井花园，一应俱全，布局考究、工艺精湛。

Xiamei Village, located in the Wuyi Mountain, is a small village that grew from a tea market. The Dangxi River traverses from west to east through the village, flowing into the Meixi River, then converging in the Chongyang River, forming a river network which facilitates the water transportation. At the beginning of the Qing Dynasty, this place was the economic and trade center of a tea market in the Wuyi Mountain, attracting non-local traders to procure here. So far, locals are still making a living on farming and tea-producing.

Steps are set along the banks of the river for bailing and washing, constituting a commercial market surrounding the water. Along the street there are houses, commercial shops, small temples, ancestral temples, hotels, workshops, verandas, wells, and so on. By the side of the river stands leisure chairs being in parallel with the riverbank or out-stretching eaves.

Of more than 30 ancient vernacular houses from the Ming and Qing dynasties in the village, there are merchant's residences, official's mansions, the scholar's school, and the representatives are: the Official Residence of the Zou Clan, the Ancestral Hall of the Zou Clan, the Hermit Home of the Cheng Clan, the Xishui Villa, the Officer's Residence of the Fang Clan and other buildings. The architecture are constructed with buildings in wood and brick structure, east pavilions and west wing-buildings, balconies and penthouses, patio gardens and so on, featuring thoughtful layout and exquisite arts and crafts.

下梅村当溪水街 The Dangxi Water Street in the Xiamei Village

下梅村梅溪景色 The scene of the Mei Stream in the Xiamei Village

水街檐廊 The rain-proof bridge

民居天井 The small yard in a vernacular house

民居入口 The entrance of the vernacular houses

民居梁架 The wooden beam in the vernacular houses

福建崇安城村

Cheng Village, Chong'an County, Fujian Province

城村位于福建武夷山市兴田镇，2000多年前的闽越王城曾在此修建。村北是古渡码头，曰"淮溪首济"。村落门楼坐北朝南，高耸于村口。穿过门楼，进入村内，鹅卵石街道两旁古民居建筑错落有致。三条主街、36条小巷蜿蜒曲折，在村中呈"井"字形纵横交错，变化有序、严谨，完整地保留了明清时期古村落风貌。

古庙宇尚存九座，尤以慈云阁、华光庙最为壮观。砖木结构的慈云阁，三殿两厢布局，华光庙则为四殿两厢布局。颇具规模的赵氏宗祠面阔三间，中为天井，后是享堂，中央供奉祖先牌位。民居大多是一进式的三合院，古朴的砖雕门楼上高悬堂匾和楹联，屋内不施油漆，保持原木本色，院中以方砖铺地，朴实却不失庄重。庙宇、宗祠、民居的马头墙突兀多姿，层层叠叠，朴实厚拙。墙上砖雕石刻，梁架木作雕饰，使建筑艺术和装饰艺术达到了完美的统一。

Cheng Village is located in Xingtian Town, Wuyishan City, Fujian Province. Two thousands years ago the city of King Yue was built here. The north of the village is an ancient ferry terminal, called "Huai Xi Shou Ji". Its gate building facing south, towers at the entrance to the village. Through the gate into the village, cobblestone streets are lined with ancient vernacular houses. Three main streets, 36 lanes meander by twists and turns in the village and form a "井" shape. This ancient village from the Ming and Qing dynasties has kept its original style. Nine ancient temples still exist, and in particular the Ciyun Pavilion and the Huaguang Temple are the most spectacular. The Ciyun Pavilion is made of brick and wood structure, having a layout of three halls and two wing rooms, while the Huaguang Temple has a layout of four halls and two wing rooms. the large-scale Zhao Ancestral Hall is in three-room breadth, with a patio in the center and leisure hall at the back and the Ancestral Memorial Tablets are worshipped in the center. Most vernacular houses are of one-hall yard enclosed in three directions. On the ancient and simple door building built of bricks, there hang plaque and couplets. To remain the original wood feature, no paint is applied within the house, while the courtyard is paved with tile floors, simple but dignified. The horse walls of the temples, ancestral temples and vernacular houses vary in style and posture, layer upon layer, thick, simple and humble. Carvings on the brick walls and wooden beams achieve a perfect unity between the architecture and decoration arts.

宅居内院 The interior yard of the residence

城村门楼 The gate-building of Cheng Village

宅居二门 The second gate of the residence

庙宇内景 The interior scene in the temple

宅居厅堂 The hall of the residence

赵氏宗祠与牌楼 The Zhao Clan Ancestral Hall and the Memorial Gateway

过街楼 The over–the–street Building

岐山公祠 The memorial temple of the revered Mr.Qishan

福建泰宁尚书第
The Former Residence of Minister, Taining County, Fujian Province

尚书第位于泰宁县城关，建于明天启年间（1621～1627年）。主体建筑坐西朝东，五幢一字排开，每幢均是三进厅堂。五幢建筑大门前设一甬道，在甬道两端设南北大门，南大门为磨砖门楼建筑，门额嵌有尚书第巨幅石匾，北门为轿厅，是三开间硬山式平房木构建筑，各幢建筑之间以封火墙间隔，中设廊门相通。

各组建筑入口均有石匾门额。自北而南第二幢为正堂，开间尺寸最大，门前甬道扩大为前院。其入口门廊最为精致，匾额、梁柱、斗栱及墙面布满石雕、砖雕、木雕装饰，雕工精美。大门两边一对高2米的抱鼓石，鼓座上浮雕双狮戏珠、云龙、花卉等，气派非凡。

气势宏伟，布局严谨，用材恰当，做工讲究，封火山墙，设计合理，雕工精细，图案丰富，可以说是尚书第建筑的特点。

The Former Residence of Minister, located in Taining County, was built during the reign of Emperor Tianqi of the Ming Dynasty (1621 ~ 1627). The main building faces the east, composed of five buildings deployed in parallel and each building has three halls. A path runs in front of five buildings and at its two ends are the north gate and south gate. The south gate is a door building constructed of grinded bricks and a huge stone plaque is embedded in the upper side of the door. The north gate is a sedan hall, a one-floor hard-mountain wooden architecture in three-room breadth. Each building is separated by firewall with corridor doors in between to link different buildings.

At the entrance of each building, there is a stone plaque in the upper side of the door. From north to south, the second building is the main hall with the largest breadth and the path in front of it is expanded to a yard. The most exquisite part of the building is the entrance veranda which features vertical plaque, column and bracket and its wall is covered with beautiful decoration of stone, brick and wood carvings. A pair of drum stones about two meters high stand on both sides of the door, the seats of which are embossed with two lions playing with pearls, dragons in cloud as well as flowers, presenting an extraordinary style.

The mansion is of magnificence, has precise layout and adopts proper materials with delicate craftsmanship. It is characterized with the fire gable, rational design, fine workmanship and rich patterns.

天井 The small yard

入口大门 The entrance of the gate

天井 The small yard

院门 The gate of the yard

厅堂 The hall

宅居室内 The interior of the residence

福建闽清宏琳厝
Honglincuo in Minqing County, Fujian Province

宏琳厝在福建闽清县坂东镇的新壶村，由药材商人黄作宾于清乾隆乙卯六十年（1795年）始建，至其子宏琳时建成。

宏琳厝外围方形，布局严谨，纵轴布置有三进。每一进之中为正厅，正厅与后厅以屏风相隔，第三进正厅设祭祖神龛。正厅两侧为书院。宏琳厝各进之间隔一横街，由过雨亭相连，过雨亭中轴对称，进出十分方便。外墙东南角与西北角还各建一间兔耳房（望哨），用来观察四周动态。

Honglincuo, located in Xinhu Village of Bandong Town, Minqing County, Fujian Province, was built in the 60th year of Emperor Qianlong in the Qing Dynasty (1795) by a medicine businessman named Huang Zuobin. The construction of the building was completed by his son Huang Honglin.

Honglincuo has a square periphery and precise layout. Its three halls are deployed along the vertical axis. In each hall, the central room is the main chamber which is separated from the back chamber by screens. The main chamber of the third hall is used to place the ancestral shrines and its two sides serve as the school. Between each independent hall there is a street in east-west orientation and all the streets are connected by rain pavilions which are of axial symmetry, making it convenient to travel in the compound. At the southeast and northwest corners of the exterior wall there are rabbit penthouses (guard post) to observe the surrounding environment.

护厝楼居 The side building

厅堂院落 The courtyard of the hall

后座厝屋 The back house

屋顶山墙 The gamble of the roof

厅堂檐廊 The eave veranda of the hall

福建尤溪厚丰村郑氏大厝

The Residence of the Zheng Clan of Houfeng Village, Youxi, Fujian Province

郑氏大厝位于尤溪县厚丰村，始建于清乾隆年间，历时10多年建成。郑氏大厝坐北朝南，主体建筑为悬山顶木结构。整个建筑群由正厝、护厝及壁舍、厢房等组成，建筑面积2800多平方米，是集闽南、客家建筑风格为一体又极富个性的闽中传统民居。

建筑群平面呈长方形，外墙封闭。内部建筑功能齐备，有相对独立的厅堂活动区、女眷生活区、宾客休闲区、生活资料储存区等。建筑结构严谨，布局合理，出入大厝有一个大门和三个小门，具有较强的防御性能。

正厝大堂前檐轩顶木雕和扶厝一、二层大厅山墙和梁架雕花，采用了工艺复杂的镂空雕手法，线条流畅，技法娴熟，堪称一绝。

The Residence of the Zheng Clan is located in Shangcun Village, Yantian Town, Xiapu County. It took nine years to complete its construction which began in the 3rd year of Emperor Xianfeng of the Qing Dynasty (1853). The residence faces south and its main building is of a three-hall overhang-roof wooden structure. The entire complex is formed by a main building, a side building, the shrine and the wing rooms, etc., covering a floor area of 2800 square meters. It is a traditional architecture in the central area of Fujian Province, which integrates the architectural styles of Fujian area and Hakka and forms a unique character.

The building has a rectangular flat with enclosed exterior walls. The inner architectures are designed for different functions, including the independent activity area in the hall, the living area for female members of the family, the guest lounge, the storage areas and so on. It has a rigid architectural structure and rational layout. At the entrance there are one big door and three small doors leading to the residence, providing strong defensive function.

The wood carving in the main hall and gables and carved beams on the first and second floors of the side building adopt sophisticated hollow carving technique which is presented with smooth lines and skillful technique.

正厝入口 The entrance of the main building

正厝堂楼 The main building

正厝前院 The front yard of the main building

正厝屋顶 The roof of the main building

宅居大门 The gate of the residence

正厝内院 The interior yard of the main building

护厝天井 The small yard of the side building

郑氏大厝正厝堂楼天井
The small yard of the residence of the Zheng Clan in front of the main building

正厝二楼檐廊
The eave veranda on the second floor in the main building

二层护厝 The two-story side building

过水廊 The cross-water veranda

正厝门厅 The foryer in the main building

福建连城培田村

Peitian Village, Liancheng County, Fujian Province

培田村位于闽西山区连城县宣和乡河源溪上游的吴家坊，已有800多年的历史，是一处保存较为完整的明清时期客家古民居建筑群。村落由学堂书院、宗庙牌坊、民居祠堂等建筑和一条千米古街、五条巷道、两条贯穿村落的水圳组成，规模宏大，布局讲究，配置得体，壮观和谐。

培田古民居群以"大夫第"、"衍庆堂"、"官厅"等为代表，是著名的"九厅十八井"式客家建筑。"大夫第"又称"继述堂"，建于1829年，历时11年建成。"衍庆堂"为明代建筑，建筑结构与"大夫第"大体相同，但门外荷塘曲径，门前石狮威镇。"官厅"因接待过往官员而称"官厅"，高墙耸立，四周封闭，设计精巧，工艺精湛，其左右花厅则专供主人休闲会友，楼下厅为学馆，楼上厅为藏书阁。民居布局尽管厅多、井多、房多，却井然有序，厅与厅之间既有通道相连，又有门户隔阻，各成单元，既利于大家族聚族而居，又不妨碍小家庭各享天伦。

Peitian Village is located at the Wujiafang on the upper stream of the Heyuan River, Xuanhe Town, Liancheng County, in the mountainous area of the west Fujian Province. With a history of over 800 years, it is a well-preserved ancient Hakka vernacular complex of the Ming and Qing dynasties. Consisting of the private schools, the ancestral monuments, the ancestral halls, an old street of one-thousand-meter long, five lanes and two rivers running through the village, the village has an exact layout and reasonable arrangement, boasting a spectacular harmony.

Peitian ancient vernacular complex is represented by the "Residence of Minister", the "Yanqing Hall" and the "Official Hall", which are famous Hakka architectures of "nine-hall and eighteen-patio" style. The "Residence of Minister", also known as the "Jishu Hall", was built in 1829 and completed 11 years later. The "Yanqing Hall" is a building of the Ming Dynasty, whose structure is similar to that of the "Residence of Minister". There is a lotus pond and the zigzag path in front of the building and solemn stone lions stand at the gate. The "Official Hall" is given such a name because it was used to receive passing-by officials. It has high walls and enclosed yard and presents delicate design and sophisticated workmanship. The flower halls on the left and the right sides are dedicated to entertaining its owner and his friends. The hall downstairs serves as the private school and the hall upstairs for storing books. Although there are a number of halls, rooms and patios, the vernacular house is arranged in an orderly manner in the layout. Halls are connected by paths but with independent doors, which makes it convenient both for the big family to live under one roof and small families to enjoy their own leisure and union.

培田敦朴堂 The Dunpu Hall in Peitian Village

厅堂前院 The front yard of the hall

进士第 The scholar's residence

祠堂入口 The entrance of the ancestral temple

院落花墙 The tracery wall of the courtyard

厅堂天井 The small yard of the hall

村口牌坊 The memorial gateway at the entrance of the village

敞厅天井 The small yard of the open hall

宅居厅堂 The hall of the residence

敞厅陈设 The furnish of the open hall

厢房木作装饰
The wooden decoration of the wing-block

湖南凤凰古城民居
The Vernacular Houses, Fenghuang, Hunan Province

湘西凤凰古城因其西南方有山如凤形而得名。自古以来这里就住着土家族和苗族的先民，至今已有三千年的历史。

凤凰古城坐落在沱江河畔，碧绿的江水从古老的城墙下蜿蜒而过，河畔上的吊脚楼房轻烟袅袅，景色轻盈秀丽极富节奏感。现今的红色砂岩城墙乃清初康熙年间所始建，古城依山傍水，城内民居纵横毗列，极有规律。

古城民居布局大体沿袭传统院落式建筑，如沈从文故居是典型的南方四合院建筑，分前后两进，穿斗式木结构，马头墙上装饰有鳌头，故居小巧别致，门窗镂花，古色古香，清静典雅。

在陈氏老宅典型院落式民居中，天井周围下有回廊，上为跑马廊。回廊右侧有木梯上楼。宅院雕花门窗甚为精美，室内装修工艺精细。

The historic town of Fenghuang County got such a name because the mountain on the southwest looks just like a phoenix (the bird is pronounced as “fenghuang” in Chinese). From the ancient times, the ancestors of the Tujia and the Miao ethnics began to live here and it has been about 3,000 years. The historic town of Fenghuang County is situated by the Tuojiang River and the virid water winds its way along the city wall through the town. Smoke is curling upward from the kitchens of the overhanging houses by the river, with a slim and graceful sight full of rhythms. The existing red sandstone wall of the town was built during Emperor Kangxi’s reign of the Qing Dynasty. The town is situated at the foot of the mountain beside a stream and the houses in it are adjacent to each other regularly.

The layout of the houses in the town mainly follow the traditional courtyard pattern. For example, Shen Congwen’s former residence is a typical four-in-one courtyard in south China, consisting of a front and a back yard, with through-jointed wood structure, and the horse-head gable wall decorated with the figure of the head of a huge legendary turtle. The architecture is small but untraditionally shaped, with ornamental engraving, having an antique, elegant and quiet flavor.

In the Chen’s Historic Courtyard-shape Residence, typical in south China, there is a winding corridor around the small yard, on the right side of which there is a wood staircase leading upstairs. The carved doors and windows in the residence are elegant and exquisite and the interior decorative craft is fine and meticulous.

沿河民居 The vernacular houses along the river

沿河民居吊脚楼 The overhanging vernacular houses along the river

宅居入口 The entrance of the residence

古城街巷 The lanes in the historic town

古城商铺 The stores in the historic town

古城大宅 The palazzo in the historic town

街巷门楼 The gate-building of the lanes

广东顺德碧江职方第

Zhifang Mansion, Bijiang Town, Shunde District, Guangdong Province

广东顺德碧江职方第共四进，包括门厅、牌坊过亭、大厅和三层的回字楼。职方第宅主苏丕文，曾任职方司员外郎，为正三品官，于清代道光二十三年（1843年）荣归故里而建造。

职方第门厅三开间，大门正中设有仪门，仪门之后为一幢砖石结构的牌门横贯在天井中，牌门额上前后各有石刻坊匾一块，朝向前门刻有"视履考祥"，朝里对着四柱大厅的一面刻着"退让明礼"，石匾上的名人墨宝和先贤哲理具有浓郁的文儒之风。

颇有特色的是牌门与厅堂之间的过亭，过亭覆盖着第二进天井，既能扩展大厅的室内空间，又确保了大厅的通风和采光，同时丰富了中轴线上建筑的形象，过亭两侧的庑廊改为接待室，檐下采用隔扇装修。

职方第大厅墙体均用水磨青砖砌筑，檐廊采用石柱，室内中心置有四根木柱，前面木柱之间以落地门罩形成室内中心空间，后面木柱置有屏门。整组建筑空间通透灵巧却不失威严。

Zhifang Mansion in Shunde, Guangdong Province is a four-hall building, including the foyer, the archway, the pavilion, the big hall and a "回"-shaped building with three stories. Its owner Su Pi-wen used to serve as a low-level official at the Ministry of Defense, a three-level rank official. He retired in glory in the 23rd year of Emperor Daoguang's reign in the Qing Dynasty and built this residence in his hometown.

The gate hall is of three-room breadth and in the middle of the main gate there is a ceremonial gate, followed by a masonry door of brick-stone structure lying in east-west orientation in the patio. Both on the front side and back side of the masonry door there is a stone plaque, whose front side is engraved with the Chinese characters "视履考祥 (looking back the past and looking into the future)" and on the back side "退让明礼 (modesty and courtesy)". The handwritings of the ancient celebrity and sages' philosophy on the stone plaque are enriched with Confucian style.

As to the architecture, the most distinctive part is the passage pavilion between the plaque door and the hall. The pavilion covers the second patio, which not only increases the hall's interior space, but also ensures ventilation and lighting of the hall. Moreover, it also enriches the image of the building on the central axis. The verandas on its two sides are changed to reception rooms and its eaves use grid fan decoration.

The walls in the big hall adopt milled bluestone brick and veranda adopts stone columns. There are four wooden pillars in the center of the place. The pillars in front create an indoor central space in the form of a French door and pillars in back are equipped with screens. The entire building is smart but full of dignity.

职方第厅堂 The main hall of the Zhifang Mansion

天井遮盖 The small yard cover

厅堂门罩 The door-casing in the main hall

四川宜宾李庄民居

Lizhuang Vernacular Houses, Yibin City, Sichuan Province

清咸丰时李庄为四川宜宾南溪县最大的场镇。李庄镇古为渔村，汉代曾在这里设驿站，由于濒临长江，故为明、清水运商贸之地。镇上酒肆茶楼，商店林立，现仍保存明、清古镇的格局和风貌，具有浓厚的地域风味。

李庄文物古迹众多，人文景观荟萃。镇内有庙宇、殿堂、楼台、戏楼、民居等建筑，类型多样，造型丰富，雕刻精细，图像生动，有较高的艺术欣赏价值。

古建筑群规模宏大，布局严谨，石板街道，幽深窄巷，街巷道路皆由条形或方形石板铺砌而成，至今完整保存着18条明清古街巷。民居四合院，封火山墙、雕花门窗，建筑较完整地体现了明、清时期川南建筑的特点。

During the reign of Emperor Xianfeng in the Qing Dynasty, Lizhuang was the largest market town in Nanxi County, Yibin, Sichuan. It used to be a fishing village in ancient time and in the Han Dynasty it was used as an official hotel. Due to its proximity to the Yangtze River, it thrived and developed into a trade and commerce place of water transportation in the Ming and Qing dynasties. The town prospers with taverns, restaurants and shops. Toady, it still preserves the pattern and style of an ancient town of Ming and Qing dynasties, permeated with a strong regional flavor.

Lizhuang boasts a great number of cultural heritage and landscape. Architectures in the town such as the temples, the halls, the stages, the towers and the vernacular houses are rich in style and structure. The fine carving and vivid image is of great artistic value.

Ancient architectural group is of large-scale and strict layout. Stone streets and deep alleys are paved with bar-shaped or square-shaped flagstones and to this date 18 ancient streets of the Ming and Qing dynasties have been preserved. The vernacular courtyards, the fire gables, the carved doors and the windows all reflect the architectural features in the Ming and Qing dynasties in south Sichuan.

古镇街巷 The lanes in the historic town

民居天井内院 The interior part of the small courtyard of a vernacular house

内院 The interior yard

大门 The gate

内院 The interior yard

云南大理白族民居
The Bai People's Vernacular Houses, Dali, Yunnan Province

白族聚居在云贵高原西部，境内江河纵横，群山矗立，中部苍山环抱洱海，景色绮丽。村镇民居有独特风格。

白族民居平面有三坊一照壁、四合五天井等形式。屋顶曲线优美柔和，屋脊有升起，外墙少开窗。外墙山尖檐下有黑白彩绘，为避大风，用特制的封檐板和面砖镶嵌组成。

大门称门楼，平面八字形，分为有厦式和无厦式。有厦门楼为三间制，三滴水式屋面，中部宽高，两边窄低，庑殿式屋顶，翼角高翘。两侧翼墙顶部砖雕，下为条石勒脚，装饰琳琅满目，色彩华丽。

照壁上有屋顶，下有墙裙。屋脊两端高翘，檐角飞起，屋面呈凹曲状。形象优美，色调淡雅。

白族民居外貌，主次分明，高低错落，有韵律感。照壁、两厦山墙面和富有个性的门楼，近望比例匀称，形象悦目，有鲜明的民族特色。

The Bai ethnics live in the western part of the Yunnan-Guizhou Plateau. Within the region they live, rivers crisscross and mountains rise straight up. In the middle part of the region, the Erhai Lake is nestling among the Cang Mountains with a gorgeous view. The climate there is mild and the rainfall is plentiful. With plenty of historic sites, the traditional vernacular architectures in the villages and towns have special characteristics.

The plane of the Bai peoples' houses has such patterns as "three-in-one-with-a-screen-wall", "four-in-one-courtyard with five small yards" and so on. The curve of the housetop is beautiful and soft. The ridge of the roof gradually rises and curls up at the ends. The roofing assumes the shape of concave curve and windows are seldom installed in the exposed wall. The part under the arris eaves of the external wall is decorated with black-and-white drawing. To evade strong winds, the yingshan roof (the Chinese gabled roof) is built with special-made fascia board inlaid with facing tiles.

The main entrance (called arch gateway), with the shape of a plane splay, is classified into long-eave type and eaveless type. A long-eave type arch gateway is generally a three-bay house, with a three-drip roofing, the middle part of which is wide and tall while the two sides narrow and low. The wudian roof (Chinese hipped roof), with the angles of wings curling up, has an elegant bearing. The top of the wing walls are decorated with carved bricks and the foot strapped with stone flags. The decoration, with splendid colors, is really a feast of the eyes.

The screen wall has a roof on its top and a dado at its lower part. The two ends of the roof ridge curl up, the angles of the eaves protruding and the roofing assuming the shape of concave curve, with elegant forms, and subdued tinge.

The appearance of the Bai people's houses makes a distinction between the primary and the secondary parts, well-proportioned with different heights, and sense of rhythms. The screen wall, the facing of the gabled walls and the arch gateway, full of individuality, are in perfect proportion, very pleasant to the eyes, and have distinct national and local features.

宅居内院 The interior yard in the residence

建筑转角 The architectural corner

白族民居入口门楼 The gate-building at the entrance of the Bai people's houses

白族石头房民居
The stone vernacular houses of the Bai People

入口门楼
The gate-building of the entrance

喜洲民居门楼
The gate-building of the vernacular houses in Xizhou County

喜洲严宅院落 The courtyard of the Yan Clan in Xizhou County

喜洲严宅入口门楼
The gate-building of the entrance of the Yan Clan in Xizhou County

喜洲严宅檐廊天花
The ceiling of the eave veranda of the residence of the Yan Clan in Xizhou County

喜洲杨宅内院照壁 The screen wall of the interior yard of the Yang Clan in Xizhou County

喜洲杨宅二层回廊 The corridor on the second floor of the Yang Clan in Xizhou County

喜洲杨宅照壁檐饰
The decorations on the screen wall of the Yang Clan in Xizhou County

喜洲杨宅门楼
The gate-building of the Yang Clan in Xizhou County

喜洲杨宅门楼装饰 The decorations of the gate-building of the Yang Clan in Xizhou County

云南丽江纳西族民居

The Naxi People's Vernacular Houses, Lijiang, Yunnan Province

纳西族分布在滇、川、藏交界地区，极大部分在云南境内，其中70%居住在丽江自治县内。本地山区和山地占90%以上，地势险要，玉龙山屹立城北，终年白雪皑皑。气候四季温凉，长春无夏，山川秀丽。

纳西族民居基本上是参照白族民居的平面形式，但照壁、门楼、围墙的尺寸较小，因用地较小或有小坡。此外，在丽江城内有玉龙雪山流下的溪水流入全城，环街串巷，使每家每户屋前屋后，甚至屋内都有流动的清水，给纳西族人民日常生活带来很大的方便，并调节了微小气候。

纳西族民居在外观造型上别具一格，略有收分的毛石或夯土外墙，檐下四周的木壁和木窗，悬山屋顶的深远出檐，宽厚的博风板和象征吉祥的悬鱼，整座建筑外观比例协调，对比恰当，装修细部和谐，给人以轻盈飘逸、雅致朴实的感觉。

The Naxi people are distributed in the boundary areas of Yunnan, Sichuan and Tibet. The majority of their houses are largely built within the border of Yunnan Province, among which 70% are located in Lijiang Autonomous County. As mountains and hills take up more than 90% of its total area, and the Yulong Mountain stands towering to the north of the county town with snow gleaming white on its top all the year round, the terrain of the county is difficult of access. But its landscape is beautiful with a cool climate just like spring in four seasons.

The Naxi people's houses are basically built according to the plane form of the Bai people's houses, but the arch-way gate and the enclosing wall are relatively small or may have a small slope. In addition, the stream running down from the Yulong Snow Mountain flows through the whole town of Lijiang, encircling all the streets and lanes, providing running water to both the front and back, even the interior of each house, bringing significant conveniences to the Naxi people's daily life, and regulating microclimate there as well.

The exterior modeling of the Naxi people's houses has a style of their own. With slightly battered rubble-stone or rammed-earth exterior wall, wood sidings and timber windows all around under the eaves, the far-reaching soffit of the xuanshan roof (Chinese overhung gable-end roof), wide and thick bargeboard and the upside-down fish that symbolizes auspice, the whole building has a coordinated proportion in outward appearance, proper contrast, and harmonious decoration details, giving people a sense of being lissome, graceful, elegant and simple.

丽江古城水街景色

The scene of the Water Street in the historic town of Lijiang

丽江古城鸟瞰 The bird's eye view of the historic town of Lijiang

民居内院 The interior yard in the vernacular houses

民居街道 The lanes in the vernacular houses

宅居门楼 The gate-building of the residence

水井——公共交往空间 The well—the space for social interactions

民居院落 The vernacular courtyards

丽江古城街巷 The lanes in the historic town of Lijiang

民居檐廊 The eave veranda of the vernacular houses

梁枋木雕 The wood carvings on the beams

夕阳照街
The street in the setting sun

丽江古城商业街
The business street in the historic town of Lijiang

铺地纹饰 The paved emblazonry

天井铺地 The small yard paving

云南建水团山村民居

Tuanshan Village vernacular Houses in Jianshui County, Yunnan Province

建水古称临安，自元代以来就是滇南政治、军事、经济、文化和交通中心，这里有元、明、清各代古建筑近百处，古桥50余座。

建水城西边的团山古村，现今保存完好的民居有张家花园、将军第、皇恩府、秀才府、司马第、保统府等民居数十所。团山民居是中原文化和边地文化完美融合的历史载体，都是合院布置的组合群体，讲求方正，中轴对称，长幼有序，尊卑有别。民居布局独特，文化内涵丰富，保持了清代和民国时期的风貌。

Jianshui was known as Lin'an in the ancient times. Since the Yuan Dynasty, it has been the political, military, economic, cultural and transportation center in south Yunnan. There are nearly 100 ancient buildings and more than 50 bridges of the Yuan, Ming and Qing dynasties.

Several scores of vernacular houses are preserved intact in Tuanshan Village on the west of the Jianshui City, such as the Garden of the Zhang clan, the General's Residence, the Mansion of Emperor's Grace, the Scholar's Residence, the Minister's Residence, the Residence of Baotong, etc. The houses in Tuanshan Village are historic carriers of the Central Plains' culture perfectly merged with the culture of the border areas. All of them are composite groups laid out with courtyards, paying attention to the design of square and upright, axial symmetry so that they are characterized with respecting seniority and the senior taking precedence over the inferior. The overall arrangement of the houses is unique with rich cultural connotations, which maintains the styles and features in the periods of the Qing Dynasty and the Republic of China.

司马第大门 The gate of the Residence of Minister

皇恩府厅堂入口
The entrance of the hall of the Mansion of Emperor's Grace

皇恩府二门
The second gate of the Mansion of Emperor's Grace

皇恩府旁院厅堂
The hall in the side courtyard of the Mansion of Emperor's Grace

皇恩府旁院外檐
The outer eaves of the side building of the Mansion of Emperor's Grace

皇恩府宅院
The residence of the Mansion of Emperor's Grace

皇恩府二楼外檐
The outer eaves on the second floor of the Mansion of Emperor's Grace

梁枋木雕装饰
The wood carving decorations on the beams

皇恩府外檐装饰
The outer-eave decorations of the Mansion of Emperor's Grace

张家花园正座厅堂 The hall of the main building of the garden of the Zhang Clan

张家花园大门 The gate of the garden of the Zhang Clan

张家花园厢房二楼入口
The entrance of the wing block on the second floor of the garden of the Zhang Clan

张家花园旁座厢房
The wing block of the side building of the garden of the Zhang Clan

张宅五间八字形大门
The five 八 -shaped gates of the residence of the Zhang Clan

门楼木雕装饰 The wood carving decorations on the gate-building

云南石屏袁嘉谷故居

The Former Residence of Yuan Jia-gu, Shiping County, Yunnan Province

袁嘉谷故居位于云南红河州石屏县南正街22号，坐西朝东，占地面积695.8平方米，建筑面积875.9平方米，是一幢清代典型的木结构四合院民居楼房。

袁嘉谷故居大门雕梁画栋，一块红底金字上书“经济特元”的直匾悬挂门顶，二重门顶“太史第”横匾，进入三重门是一幢四合院楼房，主房高二层，两边耳房为袁嘉谷生前的书房。院子中央一棵枝繁叶茂的橙树长年翠绿，夏日时为院落遮荫，使人倍感清爽凉快。

The Former Residence of Yuan Jia-gu is located at No. 22 of Nanzheng Street, Shiping County, Honghe Prefecture, Yunnan Province, facing east and covering an area of 695.8 square meters and a floor area of 875.9 square meters. It is a typical courtyard vernacular of wooden structure from the Qing Dynasty.

Its gate has carved beams and a straight plaque with a red background and the golden Chinese characters “Jing Ji Te Yuan (the Number-one Scholar in Economy)” hanging on the top of the door. On the top of the second door there hangs a horizontal board with the Chinese characters “Residence of Tai Shi (an official who holds astronomy and calendar)”. Behind the third door is a courtyard compound building. The main building has two stories and the penthouses on both sides used to serve as the studies of Yuan Jia-gu when he was alive. In the center of the courtyard grows an orange tree. The exuberant orange tree shelters the yard from heat in the hot summer days, making the whole place refreshing and cool.

袁宅厅堂 The main hall of the residence of the Yuan Clan

宅居厢房 The wing block of the residence

天井内院 The interior of the small yard

二门入口 The entrance of the second gate

云南石屏郑营陈氏故居

The Former Residence of the Chen Clan of Zhengying Village, Shiping County, Yunnan Province

郑营村位于石屏县宝秀镇，原是明朝屯军后裔建设起来的小村庄，现居住有汉、彝、白、傣、哈尼等族，共30多个姓氏，是一个典型的民族混合居住的村寨。

郑营村历史悠久，文化内涵丰富。郑营目前保留着比较完整的陈氏、郑氏和武氏宗祠以及不少修建于清末民初的民宅。

郑营村民居房屋以土木结构的四合院为主，方位皆坐南朝北。陈氏民居是民居房屋中具有代表性的木结构建筑，走马转角楼式四合院坐南朝北，建筑各个天井之间曲折回环，二楼的走廊常向外伸做有美人靠。整栋建筑的所有的门窗均没有太多的雕刻镂饰，梁柱、外檐木作漆成沉朱色。点缀的精细雕刻让人叹为观止，石榴、佛手、蝙蝠、喜鹊等吉祥图案惟妙惟肖。

Zhengying Village, located in Baoxiu Town of Shiping County, was originally a small village built by the descendants of the army men stationing there in the Ming Dynasty. Today, families from the Han, Yi, Bai, Dai, Hani and other ethnic minorities are living here, a total of more than 30 family names. It is a typical village where a mixture of ethnic groups live together.

Zhengying Village has a long history and rich culture. Today a great number of vernaculars built in the late Qing Dynasty are still well preserved, such as the ancestral halls of the Chen, the Zheng and the Wu clans.

The vernacular houses are mainly of wood structure courtyards, all facing north. Of them, the Chen-Clan vernacular house is the representative of wooden construction. It is a corridor house with tile-covered roof and tilted eaves, the courtyard facing north and patios looping around the building. The corridor on the second floor stretches outwards to create meiren kaos (banisters that win the name for beauties to lean along their waist). All the doors and windows throughout the building have little carving ornaments and beams and wooden eaves are painted vermilion. Those dotted fine carvings are breath-taking with vivid auspicious patterns such as pomegranate, bergamot, bats and magpies.

内院正座厅堂 The hall of the main building in the interior yard

首层外檐 The outer eaves on the first floor

外檐木雕 The wood carvings on the outer eaves

陈宅大门 The gate of the residence of the Chen Clan

外檐窗花 The window decorations on the outer eaves

二楼跑马廊 The Paoma corridor on the second floor

坡地民居

The vernacular architecture on hillside land

我国土地辽阔，除西北高原多山脉，华北、东北多平原外，南方大部分地区属丘陵地带。广大村镇房屋因节约用地，大部分建在丘陵地上，它的优点是：不占农田，节约耕地，节约劳力，又充分利用空间，把坡地变为有用的房屋基地。

汉族坡地民居在平面布局上都沿袭我国传统合院式居住模式。它有中轴线，但因山坡地形的关系，建筑布置较灵活、自由。它的布局方法有多种：

1．利用台地建筑法，有纵向布置和横向布置两种建造方式。纵向台地利用，即沿等高线布置建筑物，它的优点是节约土方，外观有规律地排列，很整齐。横向台地利用，即将台地作“分级”处理，如台地分级高差不大，基地分级，屋面不分级（即天平地不平），如台地分级高差较大，基地分级，屋面也分级。

2．利用坡地建筑法。第一种是依坡法，当坡度较小时，可顺坡而建，优点是利于通风、采光和排水。第二种是屋面延伸法，当坡度较大时，屋面沿着山坡披梭而下，也称披梳法，这时，在长坡屋面上通常开气窗和明瓦来解决通风和采光问题。第三种是沿坡分层筑台法，当坡度很大时才采用，也称迭级法。它结合地形，外观处理十分自然。有的地形坡度太大，除基地作迭级处理外，建筑屋面也作分级处理。

此外，还有自由法，有的山坡地很不规则，无法同一方向作迭级处理，因而，只能因地制宜，就地布局。还有一种叫支吊法，或称吊脚楼，在地势较高而又需要挑出较大的建筑时才用它，在“吊脚楼”一章中再作介绍。

坡地民居外貌由于高低错落，层次叠差，因而轮廓优美，富有艺术感染力。此外，民居采用当地材料，如木、竹、土、石，因地制宜，因材施用，使得它的外貌粗犷、朴实并含有清秀和田野之美。

China has a large territory. Apart from the northwestern plateau abundant in mountains, north China and the northeastern region in plain, the majority of the southern part of China belongs to hilly country. To save the farmland, most of the houses in the villages and towns are built on hillsides. The advantages of doing so are: both farmland and labor force can be saved by not building the houses on cultivated land; full use can be made of the spaces and the hillside land can be turned into useful house site.

The plane layout of the Han people's houses on the hillside land follow the courtyard model of the vernacular architectures of our country. It has a central axis but with a more flexible and freer arrangement due to the landform of the hillside. There are several ways of layout:

1. The building methods of the tableland are used, comprising longitudinal and horizontal arrangements. The former arranges buildings along the isoline with such merits as saving earthwork, being tidy and neat due to the regular arrangement of the outward appearance. The latter grades the table land according to its super-elevation. If the grading super-elevation of the table land is not big, then the base is graded while the roofing is not, i. e. the top of the buildings are on the same level while the bases are not. But if the grading super-elevation of the table land is relatively big, then both the base and the roofing are graded.

2. The building methods of slopes are used, including: first, the method of building according to the condition of the slope. If the falling gradient is slight, then the houses are built downhill, which is beneficial to ventilation, natural lighting and drainage. Second, the method of extending the roofing or combing is used. When the falling gradient is relatively big, the roofing will descend along the inclination of the slope and transom windows and transparent tiles are used along the long sloping roofing to deal with the problem of ventilation and natural lighting. Third, the method of building in layers along the slope is used only when the falling gradient is very big. So it is also called the method of layered grading. The outward appearance is treated very naturally according to the landform. If the falling gradient is too big, both the base and the roofing of the building will be dealt with in layers. Furthermore, there is a free method. When the landform of the slope is quite irregular and hard to deal with in layers in one direction, then adaptation to the local conditions is necessary in the arrangement of the construction. Besides, the overhanging method is also used in building what are called overhanging houses. This method can only be used when the landform is rather high and relatively huge houses are overhung, which will be described later in the chapter about overhanging houses.

The vernacular architectures on hillside are exquisite in contour and full of artistic appeal due to the well-proportioned heights and the layered congruent form of the outward appearance. In addition, the adoption of the local material in the construction of the vernacular houses such as timber, bamboo, earth, and stone with the adaption to local conditions in accordance with the quality of the material lead to the roughness, simplicity, delicate and field beauty of the houses' outward appearance.

北京川底下村民居

The Vernacular Houses of Chuandixia Village, Beijing

川底下古村位于京西古道的向阳山坡上，村宅随地形高低变化灵活布局，组合形式多种多样。宅院规模小巧，一院接一院，用地紧凑，根据坡地地形，利用砌筑挡土墙加固地基，并在这天然的地基上营建房屋，经济实用，外观自然、朴实，层层递高，丰富多变。

民居坐落在山谷北侧的缓坡上，坐北朝南。一条街道将村落分为上下两部分。民居以村北的山包为轴心，呈扇面形向下延展。古民居以清代四合院为主体，基本由正房、倒座和左右厢房围合而成，部分设有耳房、罩房。四合院的附属建筑主要有门外影壁、门内影壁、门楼、拴马桩、上马石、荆笆棚等。

房内设土炕、地炉，方砖铺地，条砖墙裙。

门和窗的窗棂装修富于变化，有工字锦、灯笼锦、大方格、龟背锦、斜棂等。民居装饰有砖雕、石雕、木雕、字画等，雕刻装饰以象征吉祥的花卉、鸟兽为主，主要部位集中于建筑的屋脊、檐口、门墩石、门窗、门簪、门罩、墙壁及影壁等处。

Chuandixia Village is located on the sunny hillside of an ancient road in the west of Beijing. Houses are deployed flexibly along the variation of terrain height, in combination of varied styles. The small houses are arranged one after another. To save land, houses are constructed on the natural foundation which is reinforced with masonry according to the terrain slope, both economical and easy to use. The natural and simple houses are arranged layer upon layer with rich variation.

Sitting on the slope of the north side of the valley, these buildings are all facing south. A street divides the village into the upper and lower parts. Taking the hill in the north of the village as the axis, all the houses extend downward in a fan shape. Most ancient vernacular houses are courtyard ones in the Qing Dynasty, enclosed by the main house, a building facing north and the wing rooms on the right and left, some with a penthouse and a cover house. The outbuildings of a courtyard include the exterior screen wall, the indoor screen wall, the door building, the horse-fastening pole, the launching-horse stone, the bush shed and so on.

Inside the room is equipped with an earth Kang, a sunken hearth, the floor paved with square bricks, and the wall skirt made of rectangular bricks.

Decoration for doors and window frames are rich and varied, being in 工 -shaped, lantern-shaped, big pane, turtle-shaped, oblique lattice and so on. Houses are decorated with brick, stone, wood carvings, calligraphy and painting, mostly in auspicious symbols, such as flowers, birds and animals. The decorations are mainly placed in areas like the building roof, the cornices, the door stones, the windows, the doors hairpin, the door covers, the walls and the screen walls, etc.

川底下村鸟瞰 The bird's eye view of the Chuandixia Village

内院 The interior yard

宅居院落 The vernacular courtyards

宅居大门 The gate of the residence

入口照壁 The screen wall at the entrance

村落街巷 The lanes in the village

川底下村民居 The vernacular houses of Chuandixia Village

浙江永嘉坡地民居

The Vernacular Houses of the Sloping Fields, Yongjia County, Zhejiang Province

浙江坡地民居多建在向阳山坡上和溪流旁，依山傍水。建筑采用穿斗、干阑式木结构体系。浙江永嘉茶园坑民居，位于大楠溪上游的一个小支流旁，两山夹一水，十几户人家，布置得像一组建筑。

浙江永嘉张溪林坑村坐落在山坡上，山脚下一小溪潺潺流过，村屋爬坡，层层相叠，地形利用得十分巧妙，俯瞰村落，像秋天里的蝴蝶栖息在一方山坡上。村头有宗祠、四合院大屋，由寨门、桥屋、庙组成的村头水口建筑十分精妙。

The vernacular houses of the sloping fields in Zhejiang Province are mostly built at the foot of a hill facing the sun or by a stream. They are built with through-jointed and crisscrossed-wood-pole structure. The houses of the Chayuankeng Village in Yongjia County are situated at a little branch of the upper Da'nan Stream, which is located between two mountains. Ten odd households are arranged with the characteristics of a group of architectures.

The Zhangxi Linkeng Village of Yongjia County is situated on the hillsides, with a little stream running at the foot of the hill. The houses in the village rise one higher than another by climbing the hill just like butterflies perching on the slope in the autumn when the village is looked down since the topography is ingeniously applied. The ancestral temple, the four-in-one courtyard big houses are located at the end of the village. And the architecture of the water gap at the end of the village, consisting of the gate of the village, the bridge houses and the temple, looks extremely exquisite.

永嘉茶园村 The Chayuan Village in Yongjia County

张溪林坑村民居 The vernacular houses in the Zhangxi Linkeng Village

永嘉屿北村凹形围宅 The 凹 -shaped encircled houses in the Yubei Village, Yongjia County

屿北村入口 The entrance of the Yubei Village

屿北村茂秀堂门楼 The gate-building of the Maoxiu Hall in the Yubei Village

屿北村某宅 The residence in the Yubei Village

屿北村石墙民居
The stone walls of the vernacular houses in the Yubei Village

屿北村闲存堂门楼
The gate-building of the Xiancun Hall in the Yubei Village

屿北村穿斗式木构民居 The wood architectures with a through-jointed frame of the vernacular houses in the Yubei Village

茶园村某宅 The residence in the Chayuan Village

张溪林坑村头寨门、桥屋
The gate of the village and the bridge houses at the end of the Zhangxi Linkeng Village

浙江永嘉黄南林坑村

The Huangnan Linkeng Village, Yongjia County, Zhejiang Province

位于"Y"形的一个深谷里，涧水潺潺，青峰滴翠。毛姓，已有700多年建村历史，都是普通耕农，没有深宅大院，房屋依山抬阶而上，层层相叠，屋基、墙裙都用卵石垒筑，院子也由石子干砌，薄屋顶，大挑梁，木板墙。虽处深山，仍可用小桥、流水、人家来形容村落的基本面貌。

The village is situated in a deep Y-typed valley with gurgling gill water and green hills around it. The surname of the villagers, average farmers, is Mao and the village has a history of more than seven centuries. There is no mansion with many courtyards and very high walls in the village and the houses there are built along steps one higher than the other. The foundations and the wainscots of the houses are both built with cobbles; the roads and courtyards are also laid dry with pebbles. The houses there are built with thin roofing, big outriggers and wood-plank walls. Although the village is situated in remote mountainous areas, the basic feature can still be described as a small bridge over the flowing stream with some cottages.

黄南林坑村穿斗式民居建筑群

The vernacular architectural complex with a through-jointed frame in the Huangnan Linkeng Village

村落鸟瞰 The bird's eye view of the village

村口廊桥 The gallery bridge at the mouth of the village

依山而建的林坑民居 The Linkeng vernacular houses built along the hillside

湖南湘西苗族山寨
The Miao People's Blockhouses, West Hunan Province

湖南湘西苗寨依山而建，峰林重叠，在峰岩下形成了高低错落的青瓦木屋。

苗族民居为穿斗式木结构建筑，除屋顶的瓦外，梁柱、墙身都用木材来做。平面三开间，中间为厅，两旁为房，厨房设在房屋的一侧，形成曲尺形。建筑根据地形起伏，将旁侧厢房做成两层楼，二层转角有阳台出挑，上做美人靠，里面是姑娘的闺房，称为"小姐楼"。弯曲的青石板路将苗寨建筑串在一起，石板路时而建筑相夹、时而拱桥横卧、时而踏级上下、时而流溪傍伴，人在其中，感到一种清新自然的山野气息。

Built at the foot of the mountains, the Miao People's houses in West Hunan Province are uneven and irregular wood ones with grey tiles as the roofing under the overlapping ridges and peaks.

The Miao People's houses are wood architectures with a through-jointed frame, whose beams and walls are all made of wood except the tiles on the top. They are generally of three-bay, the middle of which serves as hall and the two sides as bedrooms. The kitchen is arranged on one side of the house, forming a carpenter's square pattern. According to the wavy terrain, the chambers standing on both wings are built of two stories and the corners of the second floor have an overhanging-eaved veranda, on which a meiren kao (banisters that win the name for beauties to lean along their waist) is set. The inward of the second floor is used as the daughter's bedroom, named "the Miss' Building". The winding bluestone roads, sometimes laid between houses on both sides, sometimes being accumbent on an arch bridge, sometimes ascending the stairs and sometimes placed along a stream, string the architectures in the Miao People's blockhouse together. You can feel the fresh tinges of mountains and plains when you are walking on the roads.

山坡错落的民居 The vernacular houses ups and downs along the hillside

民居内院 The interior yard in the vernacular houses

苗寨全貌 The whole view of the the Miao People's blockhouses

苗族石砌寨门 The stone stockade of the Miao People

木结构苗居
The wood–frame Miao People's vernaculars house

苗寨望楼 The belvedere of the Miao People's blockhouses

山地坡道 The ramp in the mountain area

云南昆明一颗印民居

The "Single-seal" Courtyard Houses, Yunnan

在云南滇中以昆明为中心的广大地区，盛行一种两层楼的小型四合院式民居，面宽仅三间，宅基地盘方整，墙身高耸光平，远望其形如印，当地称之为"一颗印"。

典型的"一颗印"规制为"三间四耳倒八尺式"，即正房三间，厢房、耳房各两间，即四耳，门厅进深仅八尺。房屋为两层，在正耳房之间留有窄巷，安置楼梯，称作楼梯巷。楼下厅堂作日常起居用，二楼正厅作为祖堂或佛堂。耳房作卧室。大门内设屏门，以挡视线。

一颗印民居，中轴对称，楼房虽为两层，但开间进深不同，屋面有高矮，前低后高。两侧耳房屋坡向内，形成"四水归塘（堂）"。四周少开窗，或开小窗，木构架，木梯泥墙。外观朴实，比例匀称稳健。

In the vast area taking Kunming as its center in the middle part of Yunnan, prevails a kind of four-in-one courtyard with two-storey houses comprising only three rooms. The house site is square and the wall is tall and bare. Seen from a distance, it looks just like a square Chinese seal in three dimensional form. So people there call it the "single-seal" courtyard house.

A typical "single-seal" courtyard house comprises three major rooms and four chambers plus a front block, and the distance between the gate and the courtyard is only 8 chi (chi is a unit of length, which is equal to one third meter). A house has two stories and there is a narrow lane between the major rooms and the chambers, with staircases built, thus called the staircase lane. The hall on the first floor is the living room and the principal room on the second floor serves as the ancestral hall or family hall for worshipping the Buddha. The chambers are used as bedrooms. A screen is built inside the gate for keeping out the line of sight.

The central axis of the "single-seal" courtyard house is symmetric in the plane layout. The houses are of two-stories, but because the spatial depths vary and some of the roofs are taller than those of others, the frontage houses are lower than the back ones. The roofs of the chambers slope down inwardly, gathering water to the pond (hall) (in Chinese the characte"塘" (pond) has the same pronunciation of "堂" (hall)). Windows are seldom open or only the small windows are open sometimes. The houses have wood frames and mud walls and so their exterior appearance is simple but well proportioned and steady.

屋顶与天井口 The roof and the mouth of the small yard

天井内院 The interior of the small yard

宅居外观 The outward appearance of the residence

民居大门 The gate of the vernacular houses

云南泸西城子村土掌房

The Chengzi Village's Pisé Houses, Luxi County, Yunnan Province

在云南省红河哈尼族彝族自治州一带，由于天气炎热干旱，当地彝族人民居住在一种土墙、土顶、外墙无窗的房屋内，叫做“土掌房”。

城子村是最具特色的土掌房建筑的村寨，坐落在云南省泸西县永宁乡。村落依山而建，层层而上的土掌房，形成一级级台阶，前后相衔，左右毗连，从山脚一直建到山顶。

这种房屋平面三开间，近方形，二至三层。分前后两部分布置，前面是一层厢房，作为厨房、杂物间，后面为两层正房。正房楼上堆放谷物。楼下的明间是堂屋，堂屋后墙中央设置祖先堂，供奉祖先牌位。

楼面和屋顶是用泥土筑的平顶，在密肋圆木上铺树枝粘土并夯实，做成晒台，用木梯上下。大的土掌房有天井、院落，成四合院形式。

土掌房墙体厚实，有较好的隔热性能和防火性能，冬暖夏凉，取材容易，操作简单，既方便又经济。

In Honghe Autonomous Prefecture, Yunnan, because the weather is both hot and dry, the local Yi people live in the houses called “Tuzhangfang (pisé houses)”, the walls and roofing of which are made of clay and the houses have no windows in the walls.

The Chengzi Village, situated in Yongning Township, Luxi County, Yunnan Province, is a blockhouse village that possesses the most distinctive characteristics of the pisé houses. The village is built at the foot of a mountain, and the pisé houses constructed layer upon layer ascending the slope form many steps. They are joined one another from the front to the back, adjacent to each other from the right to the left, straight forward to the top from the foot of the mountain.

Such kind of approximately tetragonal houses are of three bays with two or three stories, laid out in the form of two sections from the front, which consists of one wing used as the kitchen or the lumber room, to the back, which include two principal rooms, over which grains are stored. The main room is on the first floor and in front of the back wall of the main room, the ancestors' shrines are arranged where the ancestors' memorial tablets are worshiped.

The floor and the roof of the houses are flattops made of clay, tamped on branches on the basis of rib logs to form flat roof (for drying clothes, etc.) with staircases for people to go up and down. Big pisé houses have small courtyards or compounds to form courtyard houses.

The walls of the pisé houses are very thick and solid, having rather good heat-proof and fire-proof qualities, and so it is warm in winter and cool in summer for people to live in such houses. In addition, both convenient and economic, it is easy for people there to draw materials and simple to build such kind of houses.

层叠错落的土掌房建筑群
The architectural complex of the Pisé Houses properly distributed with distinct gradation

土掌房外观 The outward appearance of the Pisé Houses

内院天井 The small yard in the interior yard

府第式楼居 The mansion-style residence

入口大门 The entrance of the gate

宅居正房 The main block of the residence

层叠错落有序的土掌房建筑群
The architectural complex of the Pisé Houses properly distributed with distinct gradation

贵州镇宁布依族石头寨

Zhenning Stone Stockade of the Buyi Ethnic Group, Guizhou Province

布依族主要分布在贵州南部、西南部和中部的高原地区，因山区地理环境的特征而形成了独特的石建筑居住形态。布依族村寨和民居的特征是：依山就势、高低参差；村寨建筑沿等高线布置；群体轮廓层次丰富；村寨环境自然粗犷。

贵州镇宁布依族石头寨位于世界上最大的喀斯特地貌地区。这里石山凸起、峰连岭复，石寨依山傍水，寨前小河环村，寨后绿树成荫，自然风光秀丽，环境优美。

进寨门后，仿佛进入了石头世界，有石砌墙壁、石板屋面、石路、石阶、石门以及家用的石磨、石槽、石桌、石凳等。石屋顺山坡建造，层层叠叠沿等高线作台阶式布置。远望山寨，石房与石山混为一体，绿树点缀其间，一片自然景象。

石屋就地取材，利用附近的水成岩，按石纹分层自然开采，每层厚度相仿，干摆，不用灰浆。石墙有平整挺拔的平毛砌，也有自然活泼的乱毛砌。墙身上有石砌尖拱或圆形窗口。屋顶俯视有铺作菱形和自然片石之分。石房两端山墙挑檐做简单粗犷的曲线，充分表现出山地民居敦厚质朴的个性。

Mainly inhabiting in southern and southwestern Guizhou and the central highland of the Province, the Buyi ethnic group has created their unique stone housing according to the geographical features of mountainous area. The village and vernacular houses of the Buyi ethnic group have the following characteristics: they are constructed according to the natural terrain with different elevations; the houses in the village are deployed along the contour line; the outline of the houses has a great variety; they enjoy original and coarse natural environment.

The Zhenning Stone Stockade of the Buyi Ethnic Group in Guizhou Province is situated in the largest area of the karst topography of the world. In this area, stone mountains bulge and ridges and peaks rise one after another. The stockade is situated at the foot of the mountains and overlooking a river, which is encircling the stockade. It has a very beautiful environment with a graceful landscape and trees shading the streets.

Entering the stockade, one will feel as if he is in a stone world: stone walls, stone-board roofings, slabstone-paved roads, stone stairs, stone doors and the domestic stone mills, stone mangers, stone tables, stone benches and so on. Stone houses are built along the slopes of the mountains, arranged on altitudes in line tier upon tier. Seen from the distance, the stone houses are identified with the stone mountains, with green trees dotted among them in a natural landscape.

The Buyi people build their stone houses by drawing on local resources and making full use of the local hydrogenic rocks. They exploit the rocks naturally according to the layers of bats. Then they just pile the slabstone up to make the wall without using any mortar because the thicknesses of the rocks are almost the same. So some of the stone walls are laid neatly and straightly while others are roughly and sprightly. The walls have stone arch-top or round-top windows. Overlooking the roofings, one can see that some are paved into the shape of diamonds while others are just paved naturally with flags. The Buyi people make some simple and rough curves on the hanging eaves on the gable walls of their stone houses, which fully display the simple and honest personality of the mountain residents.

村寨远眺 The blockade village seen from the distance far away

拾级而上的民居 The vernacular houses going up the stairs

石拱门院墙 The walls of the stone arch

民居室内 The interior part of the vernacular houses

民居内院 The interior yard of the vernacular houses

街巷拱门 The arch in the lanes

干阑民居

The vernacular houses with crisscrossed-wood structure

我国少数民族居住在西南山区，不少地方是山坡地带，当地盛产木、石，因此，他们创造了依山坡地带的半坐半悬、用木柱支撑悬挑的住屋，一般有2～3层，木构架、木楼板、木梯、干阑式构造，底层养牲畜，楼上住人，称为吊脚楼。吊脚楼前有村道与公路相通，其后建于山坡，有的山坡较陡，当地村民就用细长的木柱在山坡上支撑着大屋檐下的大木楼，呈现出既惊险又稳定的粗犷、淳朴的外貌。侗族、苗族、壮族等民居建筑中的吊脚楼就是典型的实例。

侗、苗、壮族主要分布在贵州、广西、湖南交界的地区，地势高峻，森林茂密，溪流纵横，气候温湿。他们聚族而居，有大寨、小寨，其布局无一定规则，大多选择近山傍山之处，建筑随地形自由伸展。住宅为单栋木楼，多层，户户紧密连在一起，中间只留狭长小道。但寨中必有一大块作为公共活动场所的空地。侗族寨内在这块大空地上就建有一座高大的多层立柱式建筑，称为鼓楼，这是同一村寨、同一族群在社会、政治、文化方面聚众议事的活动中心，又是文化娱乐中心，它是侗族聚落的重要标志，代表着侗族文化，称为“鼓楼文化”。

风雨桥是横跨溪上的交通建筑，既是跨河通道，又是族内居民平时休息交往的空间，节庆时也是唱歌饮酒、吹奏芦笙的地方，它位于入口处，是进寨的标志。

侗、苗、壮族民居是三层穿斗式结构干阑式木楼，当地称为吊脚楼。其上两层通常从架空层挑出，或逐层向外挑出，设外廊，屋顶为悬山顶，盖瓦或覆杉皮，坡度平缓，出檐深远。吊脚楼外形上大下小，上实下虚，装饰简朴自然，不施油饰，呈现出轻盈、生动、灵活、质朴的风貌。

Many minority nationalities of our country live in the mountainous areas of the southwest, among which many places are hillside land teemed with wood and stone and so they have built their houses along slopes that are half-seated and half-hung in the air with wood columns sustaining the overhanging parts. Such houses are called overhanging houses of two or three stories with wood frame, wood floor, wood stairs and crisscross-wood-pole structure, in which livestock are raised on the ground floor while people dwell on the second and third floor. The rear of an overhanging house is situated on a slope with a village road in front of it leading to highways. Some slopes are rather steep, so the local villagers prop up the big wood houses under big eaves with long but thin posts, presenting a breathtaking and steady outward appearance with roughness and simplicity. The overhanging houses of the Dong people, the Miao people and the Zhuang people are typical examples among such traditional vernacular architectures of China.

The nationalities of the Dong, the Miao and the Zhuang mainly dwell on the boundary areas of Guizhou, Guangxi and Hunan Provinces, where the relief of the land is high and precipitous, the forests dense and thick, with a crisscross network of rivers and streams, and a damp and hot climate. People live in big or small stockades with their kinsmen, the layout of which has no rules and most houses are built at the foot of a hill, stretching freely according to the landform. The houses are detached wood ones with several stories. They are closely connected with each other with a narrow and long walkway in between. However, there must be a big opening serving as the place for public activities. In a Dong people's stockade, a huge multi-storied upright building called the drum tower is built on the opening used as the center for the people in the stockade or of the same clan to gather together or discuss social, political and cultural affairs. As an important symbol of the Dong people's residence, the drum tower is also the center of cultural entertainment, representing the Dong people's culture, and that is why it is called "Drum-tower culture", too.

A rain-proof bridge is a transportation construction over a river or a stream, which is not only the passageway across the river, a space for inhabitants of a same clan to take a rest and communicate with each other at usual times but also the place for them to sing songs, drink alcohol and play lusheng (a reed-pipe wind instrument). It is located at the entrance of the stockade as the symbol of the entry to the stockade.

The houses of the Dong, the Miao and the Zhuang people are three-storied, crisscross-wood-pole buildings with a through-jointed frame, which are called overhanging houses by the local people. Two stories on the top are usually overhung from the sub-framing supported floor or overhung outward floor by floor. The house has a side corridor, a xuanshan roof (Chinese overhung gable-end roof) covered with tiles or fir skin, with a gentle inclination and far-reaching eaves. The overhanging houses look huge upward but small downward, the upper part being void while the lower solid. With simple and natural decorations, but without paintwork, an overhanging house presents lissome, vivid, flexible and unaffected style and features.

湖南湘西土家族民居

The Tujia People's Vernacular Houses, West Hunan Province

土家族人民多居住在湘、鄂、川、黔四省毗邻的内陆山区。他们在雨多、雾多、湿度大的山区自然条件下，因地制宜地建造了一种干阑式木构架吊脚楼民居，人住楼上，下养牲畜，以避免住所潮湿，还能防止山洪水害。

民居布局随地形和功能的需要而灵活处理，或依山、或傍水、或向阳，集居在一起，与自然环境协调统一，融人工美与自然美为一体。建筑则视山坡的陡缓，分层筑台，在台地上建房。屋脊既有平行于山坡等高线的，也有垂直于等高线的。屋顶有做成同一高度的，也有将屋顶逐级下降而成台阶状的。在陡峭的崖壁或山溪旁等复杂地形中建房，往往采取悬挑的方法，以争取使用空间。

灵巧多姿的建筑造型形成了土家族民居的独特风格。在屋檐处理上，吊脚楼采用了屋角反翘和屋面举折的结构，其外形给人以舒展向上的美感。

Most of the Tujia people live in the mountain area bordering the provinces of Hunan, Hubei, Sichuan and Guizhou provinces. Under the condition of excessive rain, fog and humidity, they build a kind of overhanging houses with wood-frame, crisscross-wood-pole structure in line with the local conditions and materials, in which livestock are raised on the ground floor while people dwell on the second and third floor so as to avoid living in damp places and protect themselves from mountain torrents.

The layout of the houses is treated flexibly based on the topography and the needs of the functions. The houses may be built at the foot of a hill or by a stream, may face the sun or built in clusters, cohere with the natural environment to fuse the artificial beauty and the natural landscape into a whole. The method of building houses in layers along the slope is used according to the falling gradient. The ridges are deployed either along the parallel or vertical contour lines. Some roofings are built with the same height while some others are built downward layer upon layer along the terrain. To build houses on complicated landforms such as by a steep cliff of a stream, the method of half-hanging the houses in the air with wood columns sustaining the overhanging parts is used to gain more space for use.

The varied and gracefully-shaped houses form the unique style of the vernacular architecture of the Tujia people. In the treatment of eaves, the overhanging houses adopt the structure of back-curl corners and arc abat-vent, displaying a beautiful and elegant taste with its contour.

湘西土家族住居山地环境
The mountainous dwelling environment of the Tujia People in West Hunan Province

庭院 The courtyard

民居二楼悬挑的姑娘阁 The daughter's cabinet of the out-stretching eaves on the second floor of the vernacular houses

凹形民居布局
The reasonable layout of the 凹 -shaped vernacular houses

湘西王村街巷
The lanes in the Wang Village in West Hunan Province

外墙 The exposed wall

贵州雷山
郎德上寨苗居
The Miao Vernacular Houses of Langdeshangzhai Village, Leishan, Guizhou Province

郎德上寨位于黔东南雷山县，是苗族聚居的村寨。村寨依山傍水，四面群山环绕，寨前有一条清澈见底的望丰河。村寨坐南朝北，建筑依山就势，疏密相间，寨内道路随地形弯曲延伸，路面用青石板或鹅卵石铺砌，整齐清洁。

寨内民居均为青瓦屋面吊脚木楼，宽敞适用。吊脚楼后半部置于基岩之上，前半部以木柱架空。民居多为两至三开间，也有四、五开间的。火塘和厨房置于后半部，利于防火。吊脚楼，底层是用于生产、饲养的吊脚层，上面是堂屋起居层和卧室阁楼层。各家都有小憩凉台和坐凳栏杆，供歇息或在此凭栏观赏自然景色。民居的前部和堂屋有垂花吊柱和门窗花饰，几何纹样不同，但都简洁朴素，给人一种清纯而粗放之美感。

Langdeshangzhai Village, where the Miao people live in a compact community, is situated in Leishan County, the southeast of Guizhou Province. The village is built at the foot of the hills and near the water, embraced by mountains all around, with the Wangfeng River in front of the village, which is so clear that you can see the bottom. It faces north, and its construction makes full use of the natural terrain with uneven density. Within the village, the roads, paved with blue slabstone or cobblestone, extend meanderingly along with the terrain, tidy and clean. The architectures in the village are all wood-frame overhanging houses covered with black tilts, which are both commodious and practical. The back half of the houses is seated on pedestal rock along slopes and the front half hung in the air with wood columns sustaining the overhanging parts. They are mostly of two or three bays and some are of four or five bays. The fire pit and the kitchen are placed in the back half, convenient for fire prevention. Livestock are raised on the ground floor while people dwell on the second and third floor. All the houses have verandas where people can take a rest or enjoy the sight of the natural landscape. In the front of the front half of the houses there are hanging posts with floral pendants and decorations on the doors and windows with different patterns of geometrical lines, which are concise and plain, displaying a sense of pure and extensive beauty.

远眺山寨 The blockade seen from the distance far away

苗寨广场 The square of the Miao People's blockade

苗寨干阑式民居建筑群
The architectural complex of the Miao People's blockades with the crisscrossed-wood-pole structure

苗居吊脚楼
The overhanging houses of the Miao People's vernacular houses

苗家姑娘 The Miao girls

村寨坡道 The ramp in the block village

寨门 The gate of the stockade

贵州黎平肇兴侗寨

The Vernacular Houses of Zhaoxing Dong Stockade Village, Liping County, Guizhou Province

肇兴大寨旧名"肇洞"，位于黔东南黎平县南部，它坐落在海拔410米的两山之间的谷地上。民居沿山谷溪流及道路两侧布置，两条溪水蜿蜒曲折于村寨中部汇合，流向西北。

大寨内分为5个团寨，5个团寨都建有各自的鼓楼、风雨桥和戏台。从高处远眺，高耸的5座鼓楼竖于村寨干阑式民居木楼之中，犹如5棵巨型银杉屹立于寨中。5座风雨桥横跨在穿寨而过的溪流河水之上，极富浓郁的侗乡风情。

鼓楼是全寨人集会议事、节目庆典之场所。其形制为：重檐层数为奇数，表示吉祥如意。建筑为木结构，穿榫衔接，不用铆钉。16根柱子分成内外两圈，内4柱，外12副柱，16根柱子托起一根不落地的冲天大柱。

Zhaoxing Dong Stockade Village, which used to be called Zhao Cave, is situated in a valley between two mountains as high as 410 meters, in the south Liping County of the southeast of Guizhou Province. The vernacular houses are built on both sides of the roads along two streams winding to meet at the center of the village and then running towards north.

The whole village is divided into five groups, all of which have their own drum tower, storm-proof bridges and stages. Seen from the distance far away, the five drum towers stand tall and erect among the overhanging houses with wood-frame, crisscross-wood-pole structure, just like five huge fir trees. The five storm-proof bridges stretch across the streams running through the village, full of the flavor of the Dong ethnic groups.

The drum tower is the place where all the people in the stockade gather together to discuss social, political and cultural affairs and hold celebrations in festivals. Its shape and structure is characterized with the odd number of double-eave layers, standing for good fortune as one wishes. The architecture is made of wood frame connected with joggle joints without any rivet. Sixteen columns are used in two circles, four in the interior circle and twelve in the exterior one, which together support a huge column shooting up to the sky.

肇兴侗寨 Zhaoxing Dong Stockade

干阑式民居
The vernacular houses with the crisscrossed-wood-pole structure

村寨鼓楼与风雨桥　The drum tower and the rain-proof bridge

鼓楼内部梁架 The interior beams of the drum-tower

鼓楼外部装饰 The exterior decorations of the drum tower

穿着盛装的侗族姑娘 The Dong Girls in splendid attire

沿河民居 The vernacular houses along the river

干阑式民居
The vernacular houses with the crisscrossed-wood-pole structure

鼓楼广场 The drum tower square

广西三江侗寨

The Dong People's Blockaded Houses, Sanjiang County, Guangxi Zhuang Autonomous Region

侗族人民聚群而居，鼓楼及鼓楼广场是聚会与交往的中心。鼓楼在侗族民间享有崇高的地位，有丰富的文化内涵和民族精神寓意。风雨桥也是侗族村寨的另一标志和象征，村民们在这里拦歌敬酒，欢送村民出寨，笑迎归来平安，多建于村寨的下游入口。鼓楼的雄浑厚重，象征着侗家的淳情古朴，风雨桥的细腻风采，体现寨民的智慧聪明。

侗族民居常依山坡等高线而建，建筑平面一般呈矩形，多坐北向南。建筑为穿斗式木构架，常为五间四架，明间小次间大，楼高二到四层不等，建筑层层出挑，通常在三、四层挑出腰檐。屋面为小青瓦悬山顶。

建筑通体木质，互相连接，不加粉饰。因山区潮湿，民居底层堆放杂物、饲养禽畜。二层以上为住居，外间是长廊，供休息及纺纱用，里间为卧室，中间是堂屋，设有火塘，全家在此就餐，冬天则休息取暖。

The Dong People live in groups, with the drum-tower and the drum-tower square as their gathering center. The drum-tower is highly valued by the Dong nationality for it has rich cultural connotations and profound messages of the national spirit. The rain-proof bridge is another sign and symbol for a Dong blockade village, on which villagers ask someone to stop to sing them a song by proposing a toast, seeing off a villager (some villagers) from the village or welcoming him (them) to come back safe and sound. A rain-proof bridge is generally built at the entrance of the lower reach of a blockade. The grand and powerful drum-tower symbolizes the Dong people's characteristics of pure sentiments and primitive simplicity while the exquisitely elegant appearance of the rain-proof bridge embodies the villagers' intelligence and wisdom.

The vernacular houses of the Dong people are mostly built on altitudes in line. The architectural plane of the houses is generally of rectangle, most of which are five-bay, four-frame ones sitting in the north and facing the south, with a through-jointed frame. The middle room is smaller than the rooms on both sides; they are of two-to-four stories, overhanging outward floor by floor, and the waist eaves are overhung from the third or fourth floor. They have overhanging gable roof made of small grey tiles.

The houses are made completely of wood, connected with each other without any whitewash. Due to the dampness in the mountains, the first floor of the houses is used to dump varia and raise fowls and livestock. Dwelling is from the second floor above. Outside the bedrooms that are inner ones with a hall in between, there is a covered corridor used for the family members to relax themselves or to spin cotton into yarn. There is a kitchen range in the hall where the whole family have their meals and warm themselves in winter.

三江侗寨 The Dong People's blockade in Sanjiang County

三江独侗乡坐龙寨
The Zuolong Blockade of the Dudong Village in Sanjiang County

风雨桥 The rain-proof bridge

鼓楼 The drum tower

广西龙胜壮族金竹寨

The Jinzhu Blockade of the Zhuang People, Longsheng County, Guangxi Autonomous Region

龙胜金竹寨 The Jinzhu Blockade in Longsheng County

金竹寨位于桂北龙胜各族自治县，是壮族干阑式民居村寨。整个寨子镶嵌在凹形的半山腰上，山上翠竹拂绿，山下溪水流淌，村寨近水傍田，享受到充沛的日照，也回避了阴湿和寒冷。

从几里之外的高山上用竹筒引来的涓涓山泉，翻山越岭，跨沟过坎，流进了山寨梯田。寨内溪水山泉在民居屋间川流，迂回曲折、时隐时现、水声叮咚，构成山寨环境的又一特色。

山道曲曲弯弯，时而拾级而上，时而穿楼过屋，富有情趣。村寨中心的大树古木参天，浓荫滴翠，构成村寨空间环境序列的高潮。

民居沿等高线自由布局，疏密相间。建筑利用自然地形，或凌空飞架，或吊脚悬挑，构成虚实相间的造型。房屋通体木质，毫无粉饰，形成朴实别致的外观。

The Jinzhu Blockade, a blockade with houses made of crisscross-wood poles, is situated in Longsheng County, the north of Guangxi Autonomous Region. The whole blockade is embedded in the concave mountainside, with emerald bamboo groves on the top, and water in the stream running at the foot of the mountain. The blockade is close to water and beside the cropland and so it enjoys sufficient sunshine and avoids damp and cold as well.

The trickling spring channeled with bamboo tubes runs into the terraced fields after crossing over hills and crests, trenches and ridges from the tall mountain several miles away. The stream and spring runs through the houses with many twists and turns, now appearing and now disappearing, with tinkling sound, and forms another special feature of the blockade's environment.

The mountain passes are twisting along, sometimes through the buildings, full of temperament and interest. The old trees in the stockade rear their heads to the sky, whose thick branches and leaves seem to blot out the sky, putting the sequence of the spatial environment of the blockade into the climax.

The houses are built on altitudes in line flexibly with uneven density. The natural landform is made full use of; some houses are built on stilts in the air, while others' overhanging parts are sustained with wood columns, forming an appearance of void-solid combination.

龙胜金竹寨 The Jinzhu Blockade in Longsheng County

干阑式民居
The vernacular houses with the crisscrossed-wood-pole structure

宅居室内 The interior of the residence

云南西双版纳傣族竹楼

The Dai People's Bamboo Houses, Xishuangbanna, Yunnan Province

我国南方气候炎热潮湿，河谷纵横，水网密布，湿地较多。云南西双版纳的傣族就位于这种湿地环境中。

由于当地气候湿热，竹木较多，傣族人民在湿地上以竹木为支架建成两层楼房，称为竹楼。进竹楼前，先登木梯，到二楼前廊，再进入厅房。楼下由木柱支撑架空，作养牲畜用，二楼为厅房，用木板间隔。宅内不供神。民居的堂屋中设火塘，用于煮饭烧茶，家人在此团聚。火塘常年不熄火，楼内无厕所。楼外为晒台，为日常生活用。

傣族民居的外貌特点，一是干阑式竹楼，二是歇山式大屋面，当地称"孔明帽"。竹楼四周建偏厦，构成重檐，可防烈日射晒。周围高大的树木使建筑置于浓荫之下，达到通风降温的目的。硕大的屋顶，黄褐色的竹墙，通透的前廊，轻巧的竹晒台和木柱林立的架空层，构成虚实、明暗、轻重对比，在热带花木的掩映下，婀娜多姿，显露出轻盈、飘逸、通透、柔美的建筑风格。

The climate in the south of our country is both hot and damp, with river valleys in length and breadth, densely covered water networks, and relatively plenty of wetland. Xishuangbanna of Yunnan Province is just situated in such an environment.

Thanks to the hot and damp climate, plenty of bamboos grow there and so the Dai people build the two-story houses with bamboos as trestle, called the bamboo houses. Before entering the bamboo house, one has to climb a wood ladder upstairs to get to the narthex of the second floor and then the drawing room. The houses are built on wood stilts, and livestock are raised downstairs while the second floor is used as rooms partitioned with planks. No god is worshiped in the house. A fire pit is arranged in the main room for cooking meals, where family members have their reunions. The fire pit keeps the fire burning all the time. There is no lavatory in the house. The outside of the house is a flat roof (for drying clothes, etc.) for daily use.

The characteristics of the Dai people's vernacular are: on the one hand, they are crisscross bamboo houses and on the other hand, each house has a xieshan-typed big roofing, which the local people call Kongming's Cap (Kongming was the style name of Zhuge Liang, a statesman and strategist, prime minister of the Kingdom of Shu in the period of Three Kingdoms). Around a bamboo house, an outbuilding is constructed so as to form double eaves to prevent the strong sunshine. The tall trees surrounding the house put it under the thick branches and leaves with the purpose of ventilation and cooling. The huge roofing of the house, the yellowish-brown bamboo walls, the permeable front veranda, the light and handy bamboo flat roof and the aerial stratum with wood columns in great numbers, display a slim and graceful, elegant, permeable, gentle and pretty architectural style against the tropical flowers and wood with a gracefully slender figure, in comparison with void and solid, light and shade, triviality and importance.

傣族竹楼 The Dai People's bamboo houses

傣族村寨 The Dai People's village

水乡民居

Houses in the region of rivers and lakes

我国南方地区，江湖横贯，溪流密布，众多乡村墟镇沿河傍水而建。水乡村镇布局中，一类是乡村，以住为主，另一类是商住综合，其特点：河水贯通村内，也有沿村或环村而过。墟镇中常有两条河道汇合，方便了水乡的交通和商业。有些店铺就设在沿河的通街上，从水陆两路进店都可购物。

墟镇街巷，老百姓的住房——民居建筑，宁静而有序地整齐排列着，屋前有道路，屋后有水路，出门即可登船，水陆交通方便是水乡的一大特色，船也就成为了水乡住户不可缺少的交通工具。乘着小船在水巷中悠悠绕行，足以体会和眷恋水乡生活的无限乐趣和情感。

在村镇中，桥是水乡河道中联系各地各角的不可缺少的交通元件。它纵横相连，像蜘蛛织网，成为了水乡村镇的美好景观，如直桥、平桥、拱廊、廊桥、圆孔桥、梯级桥、宝带桥等，各式各样的桥梁，为水乡、村镇增添了无限的美好景观。此外，还在桥头建造了一些楼阁或休息亭榭，有的在河边建起了商铺食肆，形成了水乡建筑的一片繁荣景象。

远望水乡岸边，毗连的黑漆大门、灰瓦白墙的宅第住家，坐落在石板小路旁，河畔种植着杨柳、果树，其后为一片埠头，几叶小舟荡漾在小河面上，形成了一幅水乡民居的安逸幽静画面和人间休闲之美。

The south of China has plenty of rivers, lakes and streams and numerous villages and towns are built along a river or beside a stream. The layout of the region of rivers and lakes consists of two main types: one type is the village that mainly functions as the residence while the other is the complex that functions both as a trading center and a residence. A river may link up the whole village, run through it or even surround it before running away. Frequently, two rivers may join each other in a village or a town so that the transportation and trade may become more convenient. Some shops or stores are just set up on the streets along the river, which have an easy access for shopping from both the land and water.

Villages and towns, the ordinary people's houses — the traditional vernacular architectures are arranged quietly, neatly and in good order. There are roads or streets in front of the houses and waterways behind them. Boarding on boats immediately after leaving one's home is a special feature indicating the convenient transportation in the region of rivers and lakes. Therefore boats have become indispensible means of transportation in such regions. Passing round the water lanes by boat is enough to experience and be attached to the boundless pleasure and appeal of the life in the region of rivers and lakes.

In a village, a bridge is a necessary transport component to link different parts over the stream-way there. It looks like a cobweb by linking in length and breadth as a fine landscape of the village. Bridges such as straight bridges, level bridges, arch bridges, gallery bridges, circular-orificc bridges, stair-step bridges and treasury-belt bridges, and so on, add limitless fine landscape to the villages and towns in the region of rivers and lakes. In addition, some towers or pavilions are also built at the head of the bridges where people can take a rest while some shops, stores or restaurants are built beside the rivers, presenting a flourishing sight of the region's construction.

Seen from a distance, the adjoining black-paint gates, the homesteads with grey tiles and white wall situated along the slab-stone paths, along with the poplar trees, the willows trees, the fruit trees and the bending weeping willows, the stretch of wharfs behind them, several small boats drifting on the water of the stream, look together really like an easy, comfortable, peaceful and secluded picture in which people in the region of rivers and lakes can live a life of leisure.

江苏吴江同里水乡

Tongli Town, the Region of Rivers and Lakes, Wujiang, Jiangsu Province

同里镇位于江苏吴江市东北。镇内风景优美，镇外四面环水。镇区被15条小河分隔成7个小岛，而49座古桥又将小岛串为一个整体。镇内街巷逶迤，河道纵横，家家临水，户户通舟，巷内深邃，幽静宜人，屋瓦连绵，白墙花窗，具有独特的水乡风貌。

同里因优美的环境、便利的交通和丰富的物产而成为士绅退隐颐养之地，因此镇上多深宅大院和精良民居。全镇现存明代建筑十余处；退思园等清代建筑数十处，被称为明清建筑博物馆。

退思园系清代一座十分精致而别具匠心的私家花园，其布局自西往东，为宅、庭、园，横向展开。退思园于江南园林中具贴水园之特例。山、亭、馆、廊、轩、榭皆紧贴水面，园如同浮于水上。水是万物生命之源，临水、跨水、卧水、与水亲近，这是人们生活中最惬意的享受。

Tongli Town is situated in northwest Wujiang County, Jiangsu Province. Surrounded with water all around, the town has beautiful scenery. The whole town is divided into seven small islands by fifteen streams and forty-nine historic bridges fuse the seven islands into an organic whole. The streets and alleys in the town meander within the town and rivers crisscross. Every house is close to water and every home is navigable. With deep lanes and quiet environment, the tilt-covered houses with white walls and colorful windows roll uninterrupted, and have the unique style and features of the region of rivers and lakes.

The beautiful environment, convenient transportation and rich produces in Tongli Town become the reasons for the gentry to go and live in seclusion and take care of themselves there. Therefore, there are plenty of compounds of connecting courtyards, each surrounded by dwelling quarters as well as excellent vernacular houses in the town. There are more than ten historic architectures built during the Ming Dynasty and several scores of houses built in the Qing Dynasty (such as the Tuisiyuan Garden) existing in the town, which are called the architectural museum of the Ming and Qing Dynasties.

The Tuisiyuan Garden is a very exquisite private garden built with originality in the Qing Dynasty. It spreads as residences, front yards and gardens from the west to the east. The Tuisiyuan Garden is a special case among the gardens pressing close to water in the region south of the Yangtze River, in which the hills, pavilions, galleries, corridors and verandas all press to the water so that the Garden looks as if floating on the water. Water is the source of all lives and people are pleased to enjoy living close to water, striding over water, and being on intimate terms with water.

同里退思园 The Tuisiyuan Garden in Tongli Town

退思草堂 The Tuisi Humble Cottage

耕乐堂院落　The courtyard of the Gengle Building

水埠　The wharf

退思园水景　The water scene of the Tuisiyuan Garden

嘉荫堂 The Jiayin Building

水街　The Water Street

江苏昆山周庄水乡

The Region of Rivers of Zhouzhuang Town, Kunshan City, Jiangsu Province

周庄在宋代就是人丁繁盛的镇市。它四面环水，南北水面形成井字形格局。沿河两侧顺延成8条长街，房屋傍水而筑。河上有桥，街的两旁店铺林立，街紧挨着河，街道上空还有过街楼。

沈厅位于富安桥东南侧的南市街上，坐东朝西，七进五门楼。它由三部分组成，前部为水墙门、水埠，是停靠船只的码头，中部为靠街楼、茶厅、正厅，后部为大堂楼、小堂楼、后厅屋，是生活起居之处，前后楼之间均用过楼或过道阁相连，形成一个大的走马楼，厅房宏大，梁柱粗实，刻有花饰。面对正厅有精细的砖雕门楼。

周庄全镇以河成街，桥街相连，依河筑屋，深宅大院、重脊高檐，过街骑楼、古色古香，水镇一体，呈现出一派古朴、明洁、幽静的风光。

Zhouzhuang began to be a town with a flourishing population during the Song Dynasty. Embosomed in water, the town forms a pattern resembling the Chinese character of " 井 " with the surface of the river and its branches from northwest to southeast. There are eight long streets along the river branches and the houses are built on both sides. There are bridges over the river branches and stores and shops stand in great numbers on each side of the streets. The streets are close to the river branches and some buildings are built over the streets just like towers over the city gates.

The Shen's Residence, a seven-yard five-arched-gateway homestead, is located on the Nanshi Street southeast to the Fu'an Bridge, facing the west. The residence consists of three parts: the front part is a ferry for boats to berth, also serving as an exit; the middle part comprises the close-to-the-street building, the tea booth, and the main hall; and the rear part includes the Datang (major) Building and the Xiaotang (minor) Building and the opisthodome building, the dwelling quarters. All the buildings are connected with passage-buildings or corridor-pavilions to form a bigger trotting-horse building. The halls and houses are spacious and the beam columns are thick and solid, with ornamental designs on them. Facing the main hall stands the arch gateway with meticulous brick carvings on it.

The whole town of Zhouzhuang takes the river branches as streets along which houses are constructed and bridges and streets are connected with each other. With an antique flavor, the town has many compounds of connecting courtyards, each surrounded by dwelling quarters, all having double ridges and tall eaves, and buildings with covered sidewalks. Water and the town are integrated, presenting a scene of primitive simplicity, clarity and cleanness, peace and seclusion.

水埠　The wharf

水乡民居　The vernacular houses in the Region of Rivers

拱桥 The arch-bridge

厅堂 The hall

周庄街巷 The lanes in Zhouzhuang Town

浙江桐乡
乌镇水乡
The Vernacular Houses of Wuzhen Town, the Region of Rivers and Lakes, Tongxiang County, Zhejiang Province

乌镇位于浙江省桐乡县，因本地土质深黛而肥沃，遂以乌名之。乌镇是一座保存完整的江南水乡古镇，河道纵横，四条老街呈十字交叉，构成河街并行、水陆相邻的双棋盘式格局，体现出江南以水建市的特点。

古镇老街路面都由长条石板铺成，两侧为商铺，沿河的商铺只有一进，下店上宅，另一侧为前店后宅。沿街店面装饰，横梁常雕有花饰，各家都不相同，成为了乌镇老街一大特色。

乌镇依河而建的民居与河上的石桥共同构成了小桥、流水、古宅的江南古镇风韵。乌镇民居的另一大特色是临河的吊脚水阁楼，它挑出河沿，下部以木柱或石柱支撑，充分占领水面以节约陆上土地。另外，近河住户多备有小船，在住房上搭起水阁，屋下就留有了泊船的地方，水阁悬空于水面之上，显得格外轻巧、空透。

Wuzhen Town is situated in Tongxiang County, Zhejiang Province. It is named Wuzhen (the character of "Wu" in Chinese means black) because the soil in the town is both black in color and fertile. It is a well-preserved historic town in the region south of the Yangtze River, with crisscrossed streams and rivers. Four old streets form a cross and the rivers run parallel with the streets just like two checkerboards adjacent to each other, reflecting the characteristic of making construction by water in the region south of the Yangtze River.

All the old streets in the historic town are paved with slabstone lined with shops on both sides. The shops on the riverside are of only one row, with shops on the first floor and bedrooms on the second floor. The shops on the other side of the road are of yards, with shops in the front and bedrooms in the rear. The frontages of the shops along the roads are decorated differently but the horizontal beams are often carved with floridity, which is a major feature of the historic town.

The houses built along the river in Wuzhen and the stone bridge together form the graceful bearing of water flowing beneath a little bridge bear historic houses in the historic town in the region south of the Yangtze River. Another important feature of Wuzhen's vernacular houses is the water penthouses close to the river with standing pillars, which stick out over the river bank, supported with wood or stone columns, to save the land by taking advantage of the space over the water. Moreover, most of the families living by the river have boats of their own. They built their water penthouses so as to leave room for their boats to lie at anchors. The water penthouses are suspended in midair, looking extremely light and handy.

水乡沿河民居

The vernacular houses along the river in the Region of Rivers and Lakes

沿河街道 The streets along the river

大型水埠码头　The large wharf

厅堂外檐屏门　The screen of the outer eaves in the hall

跨河拱桥　The arch as the passageway across the river

石拱桥　The stone arch-bridge

商业小街　The business street

民居室内一角　One corner of the interior of a vernacular house

沿河廊街美人靠　The meiren kao in the Lang Street along the river

沿河檐廊与拱桥　The eave veranda and the arch along the river

沿河民居　The vernacular houses along the river

二楼望窗外门檐小景
The view of the outer eaves seen from the window on the second floor

砖雕门楼　The gate-building of the brick carvings

浙江湖州
南浔水乡
The Region of Rivers of Nanxun, Huzhou City, Zhejiang Province

南浔镇位于太湖之南，与苏州紧邻，有700多年历史。南浔市河穿镇而过，石拱桥、石驳岸、小街水巷、依水民居，充满着江南水乡的诗画神韵。

张石铭故居位于南栅南大街，前临浔溪河，后依鹧鸪溪，建于清光绪年间。大宅有大厅三进和西式楼房50多间。正门为轿厅，二进正厅懿德堂，后为堂楼，称女厅，楼上供女眷居住。三进为内厅，因两侧廊庑窗棂嵌石刻蕉叶，形态逼真，雕工精良，故称芭蕉厅。

东大街以东为张静江故居和以百间楼为主的民居群。东大街原是南浔古镇的第一商业街，街南即为市河，街两侧有五福楼、大庆楼、天云楼、长兴馆等一批百年老店。百间楼沿河而建，空间富有层次，形成一幅"小桥、流水、人家"的美丽景观。

With a history of more than 700 years, the township of Nanxun is located on the south of the Lake Tai, bordering Suzhou. With the stone arch bridges, stone barges, the small streets and water lanes, and the vernacular houses by the river, the town, through which the Nanxunshi River runs, is full of romantic charm of the poems and pictures of the region south of the Yangtze River.

Built during the period of the reign of Emperor Guangxu of the Qing Dynasty, Zhang Shiming's former residence is situated at the Nanshan South Street, with the Xunxi River in front and the Zhegu Stream at the back of it. The palazzo has three big halls and over 50 bays of western-style buildings. The main entrance is a sedan hall and the main building in the second yard is the Yide Hall and its rear building is the central hall, called the ladies' hall, used for the dwelling of the womenfolk of the family. The building in the third courtyard is the inner hall, called the Palm-leaf Hall because both sides of the windows of the veranda are embedded with stone bars on which the palm leaves are engraved with lifelike figures and excellent craftsmanship of the carvers.

The east to the East Street is a residential complex, taking Zhang Jingjiang's former residence and the One-hundred-room Building as the main architectures. The East Street used to be the first business street in the ancient Nanxun Town. The Shi River is on the south of the street, along whose sides stand a batch of shops or stores with a history of more than a hundred years, such as the Wufu Building, the Daqing Building, the Tianyun Building, and the Changxing Shop. Built along the river, with a rich hierarchical structure of the space, the One-hundred-room Building forms a beautiful picture of a small bridge over the flowing stream with some cottages.

百间楼沿河民居　The vernacular houses along the river of the One-hundred-room Building

沿河民居　The vernacular houses along the river

崇德堂二厅　The second hall of the Chongde Hall

河道拱桥　The arch bridge on the river

民居回廊　The veranda in the vernacular houses

石拱桥　The stone-arch bridge

芭蕉厅檐廊　The eave veranda of the Palm-leaf Hall

厅堂槅扇　The partition board in the hall

宅居厅堂　The hall of the residence

百间楼外观　The outward appearance of the One-hundred-room Building

浙江嘉善
西塘水乡
The Region of Rivers of Xitang, Jiashan County, Zhejiang Province

西塘古镇位于浙江省嘉善县，古镇依河而建，主要的十字河道成为全镇的骨架，其他河道都交汇于这两条主河。

西塘以桥多、弄多、廊棚多闻名。从高耸的拱桥上眺望，沿河绵长的长廊和屋檐下的廊柱有节奏地排列，对岸高翘的马头墙、临河入水的水埠头和平直的廊棚形成虚实、高低、黑白的节奏和对比。

江南古镇的街市，沿街店铺门前常搭有棚布，使商家和顾客免受雨淋日晒之苦，后来就做成了伸出街沿的固定的廊棚，连在一起就变成了沿街长廊。再后，全镇主要街道串联在一起，就形成了西塘全镇廊棚绵延的景色。

古镇多小巷，俗称弄堂。西塘的弄堂有120多条，按其功能分为三类：一是宅内弄，又称陪弄。二是水弄，前通街道，后通河道。还有一类弄堂用于连接两条平行的街道。

The historic town of Xitang is situated in Jiashan County, Zhejiang Province, built by the Cross River course, which is the skeleton of the whole town since it is the confluence of all the other river courses.

Xitang boasts itself for its multitude of bridges, lanes and covered corridors. Overlooked from the high-rise arch bridge, the long covered corridors by the river and the columns under the eaves are arranged rhythmically, forming a distinct rhythm and comparison of void and solid, different heights, and black and white with the upward-curled gable walls, the wharf and the straight veranda canopies.

Canopy cloth used to be hung in front of the shops or stores along the streets of the historic towns south of the Yangtze River, so that the customers could avoid being exposed to the sun and rain. Later the canopies were changed into fixed corridor canopies extruding to the streets, which became the covered corridors along the streets by joining each other.

Historic towns are rich in small lanes, called Nongtang (alleyways). Xitang has more than 120 alleyways, divided into three types based on their functions: one type is the alleyway within a homestead, also called Peiong (secondary alleyway). Another type is the alleyway leading to a street with one end and to the water course with another. And the last type is the alleyway that connects two parallel streets.

宅居内院

The interior yard of the residence

双桥相映　Two bridges contrasted with each other

西塘水乡民居　The vernacular houses in the Region of Rivers of Xitang

前院二门　The second gate of the front yard

上海青浦朱家角民居

The Zhujiajiao Vernacular Houses, Qingpu District, Shanghai

朱家角镇位于上海市西郊，北连昆山、南接嘉兴。境内地势平坦，四周湖荡密布，漕港河穿镇而过，水陆交通便捷。朱家角历史悠久，民风淳朴，文化积淀深厚，自明朝万历以来商贾云集，百业兴旺，是上海保存最完好的江南水乡古镇。

朱家角历来风光优美，文人雅士和官宦之家相继迁居于此，建宅修园。东湖街的席宅是明代老宅，正厅高大宽敞，厅柱石墩硕大厚实。课植园位于古镇北首西井街，为镇上最大的庄园式园林建筑。

古街幽巷，是朱家角的独特景色，全镇长短大小不一的古巷约有几十条，有时就在两幢老房的高墙间隐藏着一条小巷，显得十分神秘。石板老街，黛瓦民宅，深巷窄弄，前街后河，密布于古镇小巷的古老宅子，依旧散发着江南古韵之淡淡清香。

Zhujia Town is situated in the west suburbs of Shanghai, bordering Kunshan City on the north and Jiaxing on the south. The terrain within the borders of Zhujia Town is flat and the land and water transportation is convenient with lakes around the town and the Caogang River passing through it. With a long history, unsophisticated folk customs and deep cultural sedimentary accretion, all businesses are prosperous and merchants are gathering in Zhujia Town, the best preserved historic water town south of the Yangtze River since the Wanli period of the Ming Dynasty.

Zhujia Town has a wonder sight, and refined scholars and families that produced public officials for several generations moved here in succession and built their houses and gardens. The Xi's former residence in the Donghu Street was built in the Ming Dynasty. Its main entrance is big and commodious and the pier bases of the columns in the hall are large, thick and solid. The Kezhiyuan Garden, the largest manor-type garden house, is located in Xijing Street, the northern part of the historic town.

The ancientry streets and secluded lanes are Zhujia Town's unique sights. There are several scores of ancientry streets and lanes, either long or short, and occasionally a small lane is hidden between the high walls of two old houses, imparting a mysterious look to it. The stone-paved ancientry streets and the vernacular houses covered with dark-grey tiles, with streets in front of them and a river behind, and the densely-scattered old residences of the historic town still give off delicate fragrance of the graceful bearing south of the Yangtze River.

沿河民居　The vernacular houses along the river

拱桥　The arch bridge

廊桥　The gallery bridge

街巷　The lanes

“惠民桥”廊桥 The Huimin Gallery Bridge

窑洞民居

Cave dwellings

我国中原黄河流域中游地跨甘肃、陕西、山西、河南等省，有得天独厚的黄土资源。它土质均匀，分布连续，土层厚达50～200米。它的主要成分为石英构成的粉砂，颗粒较细，粘度较高，粘聚力和抗剪度较强，便于挖掘施工，既防寒，又保暖，是当地老百姓就地取材、因材施用最合适的一种居住方式，称为窑洞民居。

窑洞民居根据地形土质分为三类：

第一类是直接依山靠崖挖掘横洞成窑，称为靠山窑。再细分可分两种，一是靠山式，窑前比较开阔，民居沿坡开挖，通常底层屋面是上层窑居的平台。另一种称为沿沟式，窑洞分布在沿沟两岸的崖壁土层。沟谷较窄，窑前不开阔，但因沟窄可使两岸窑洞避风沙。

第二类是下沉式窑，也称天井窑，在没有条件做靠崖窑的平坦地带，只能在地下挖出四壁闭合的下沉院，然后再向四壁挖窑，称为天井窑。河南称之为“天井院”，甘肃称之为“洞子院”、山西称之为“地窑院”、“地坑院”。它的上下交通靠梯道。

第三类是覆土窑，也称锢窑或独立式窑洞，它实质上是一种以土坯或砖石建造的拱形房屋，上部覆土夯实。

窑洞单体平面为带拱券顶的长方形房间，相邻两窑洞中间可打通。窑洞正面为出入门口，称为窑脸，是窑洞的主要正面，下为槛窗，上为拱形窗户，其旁设出入大门。有的窑洞开窗不开门，通常作为坑床。窑洞民居外观简洁朴实，一般不显露，有的在门框、门楣作一些装饰，有的在窗户玻璃上加贴剪纸。它与黄土大地融合在一起，保持自然生态环境风貌，给人一种淳朴、粗犷的农村自然之美。

The Yellow River Valley in the Central Plains of China cuts across Gansu, Shaanxi, Shanxi, and Henan provinces with particularly favorable natural conditions of the loess resources, the geography of which is well-distributed and spread in succession with a thickness of 50 to 200 meters. As the major ingredient of the loess is silt made up of quartz whose granules are relatively fine with a rather high viscosity, quite a strong stickiness and shear stiffness and so it is easy to dig for construction. The cave dwelling, both cold-proof and warm, is one of the most suitable means of inhabitancy for the local ordinary people to draw on local resources for practical purposes.

The cave dwellings can be classified into three categories according to landforms. The first are hill-backed cave dwellings, which are dug horizontally directly into the hill or cliff. This category has two subdivisions: one is the hill-back type, in front of which there is a wide open space. The cave dwelling dug along the slope and the ground floor is also the platform of the upper cave dwelling. The other are called gully-type cave dwellings, which are distributed in the solum of the precipices along both sides of a gully. Because the gully is relatively narrow, the open space in front of the cave dwellings is small. However, due to the narrowness of the gully, the caves can easily avert wind and sand blow.

The second category are subsiding-type cave dwellings or small-courtyard cave dwellings. In the flat region where there is no condition for people to dig cave dwellings into the precipices, they have to dig down into the ground so as to build a closed subsiding courtyard with precipices on four sides and then make cave dwellings by digging into the precipices, which are called small-courtyard cave dwellings. Such cave dwellings are called "small courtyards" in Henan Province, "grotto courtyards" in Gansu Province, "basement courtyards" or "pit courtyards" in Shanxi Province, the passageway for the dwellers to come in and go out is the ladder.

The third category are the covering caves, also called detached cave dwellings, which in essence are a kind of arch buildings made of sun-dried mud bricks or masonry blocks, the upper part of which is tamped with earthing.

The mono-plane of a cave dwelling is a rectangular room with a vault on top and the two adjoining caves are opened up. The door is set in the frontage, namely, the face of the cave. The lower part of the frontage is the sill wall window while the upper part is the arch window beside which the gate is usually set. Some cave dwellings have windows but don't have doors, which are used as the pit beds. The outward appearance of a cave dwelling is concise and simple without making a display of it. Some decorations may be made on the door frame or lintel and window panes may have paper-cuts stuck on them. Fused together with the loess earth, the cave dwellings maintain the natural ecotope style and features, leaving people a sense of natural beauty in the countryside with simplicity and roughness.

陕西米脂姜氏庄园

The Jiang's Manor, Mizhi County, Shaanxi Province

姜耀祖庄园位于陕西省米脂县刘家峁村，始建于清末，是一座以窑洞为主的民居建筑群，修建在陡峭的峁顶上，外围筑以18米高的城墙，东北角设有角楼，城垣上设有碉堡，南侧为拱形的庄园堡门和曲折的隧道。民居建造顺应山形地势，总体布局构思巧妙，整个庄园融于自然，宛如天成之趣。

建筑群由上、中、下三套窑洞庭院组成。穿过拱形堡门，爬越陡峭蹬道，再过月洞门、垂花门，进入雕琢精美的窑洞四合院时，顿觉豁然开朗，别有洞天。这种层层跌落的庭院，造型优美的城堡，收放得兼的构思与古朴苍劲的意境，使人心旷神怡。

Originally built in 1874, the thirteenth year of Emperor Tongzhi's reign of the Qing Dynasty, the Jiang Yaozu's Manor is located in the Liujiamao Village, Mizhi County, Shaanxi Province, which is a vernacular complex, taking cave dwellings as its main architecture. Built on the top of the loess hills, the manor is just like a castle with 18-meter tall walls surrounding it. A watchtower stands at the northeastern corner of the castle wall and a blockhouse was constructed on the wall. The arched entrance of the manor with a tortuous tunnel is on the southern side. In accordance with the chevron and the topography, the general layout of the vernacular houses is ingeniously conceived. The whole manor integrates itself into the nature as if made in heaven.

The whole architectural complex comprises three parts from the foot up to the top of the hills, forming the lower, middle and the upper courtyards with cave dwellings. When you pass the arched entrance of the complex, climb up the cliffy rocky path, go through the moon-shaped gate, then the floral-pendant gate, and enter the four-in-one cave courtyard which is cut and polished beautifully, you will suddenly see the light and the hidden wonders. The layer-upon-layer courtyards, the exquisitely constructed castle, the conception of merging the application of both retraction and release, the simple and unsophisticated, old and strong prospect, make you completely relaxed and joyful.

中院厢房 The wing block in the middle courtyard

姜氏庄园下院 The lower house of the Mansion of the Jiang Clan

上院大门 The gate of the upper house

上院大门 The gate of the upper house

姜氏庄园中院仪门 The ceremonial gate in the middle courtyard of the Mansion of the Jiang Clan

窑居 The cave dwelling

窑居 The cave dwelling

大门抱鼓石 The drum stone at the gate

陕西米脂杨家沟骥村古寨

The Historic Ji Village Blockade, Yangjiagou Township, Mizhi County, Shaanxi Province

杨家沟骥村古寨是米脂县清同治六年所建的窑洞式庄园。

建筑群坐落在主沟与支沟环抱的山地梁峁上，寨门朝东，经涵道、蹬道，再分南北两路进入各宅院。最后，爬一陡坡才到峁顶的祠堂，从祠堂向南俯视崖下，老院和新院的窑顶尽收眼底。

老院是典型的靠山窑洞庭院，用门楼、围墙和下房组合成两进四合院。新院位于古寨的西南角，背靠30米高的崖壁，新院中毗连的11孔石窑面南背北、居高临下。

窑洞建筑群在选址、理水、削崖到利用高低错落的地貌争得良好的方位等方面，都处理得非常自然和谐和协调。

The historic Ji Village Blockade in Yangjiagou Township, Mizhi County is a cave dwelling complex built in the sixth year of Emperor Tongzhi's reign (1867) of the Qing Dynasty. The complex is situated on the loesshill surrounded with the main trench and its branches, with its gate facing east. People can enter each courtyard via either the north road or the sourth road after passing thc culvert and the cliffy rocky path. Finally, one can arrive at the ancestral hall on the top of the loesshill after climbing a steep slope. Seen southward from the ancestral hall to the places below the cliff, one will be able to capture the roofing of both the old and the new courtyards.

The old yard is a typical manor made of caves with the hill at the back in which the two-row four-in-one courtyards are formed with the archway gate, enclosed walls and the servants' lodgings. The new yard is situated at the southwest corner of the stockade, with its back against a 30-meter precipice. The 11 stone cave dwellings adjacent to the new yard faces south, built in a original technique, occupying a commanding height.

The cave dwelling complex was treated very naturally and harmoniously not only in the aspect of site selection, the precipice chopping, but also in the aspect of making use of the topography with different heights so as to achieve the best orientation of the cave complex.

窑洞民居 The cave dwellings

依山而建的窑洞 The cave dwellings built at the foot of a mountain

河南巩义 康百万庄园

The Kang Baiwan's Manor, Gongyi City, Henan Province

康百万庄园面对伊洛河，背靠邙山岭，建筑坐北朝南，环境幽美。它始建于清道光年间，至宣统年间完工，历时数十年。整个庄园包括祠堂、金谷寨主宅院、普通宅院、作坊、栈房等，是我国北方黄土高原区典型的城堡窑洞庭院住宅。

金谷寨主宅院由四个并列的四合院和一个偏窑崖院组成，除第一院设有堂屋正房外，其他院落都以砖砌崖窑作为正房。头院内，房屋纵轴对称、错落有致，第二、三、四院都设有垂花、卷棚二门，二门前有4米宽的东西走道，后院还有2米宽的通道横贯，形成了一个既相互联系又相互独立的大型宅第。

康百万庄园富丽豪华，砖木结构精巧别致，叠脊山墙气宇轩昂，门窗棂花剔透玲珑，雕梁画栋风雅华贵，家具陈设典雅古朴，给庄园增添了浓郁的地方特色。

The Kang Baiwan's Manor faces the Yiluo River on the south with its back against the Mangshan Mountain, having a very peaceful and secluded environment. Its construction began during Emperor Daoguang's reign and was completed during Emperor Xuantong's reign of the Qing Dynasty, lasting for a period of several decades of years. The manor, a typical castle-cave residence in the area of the Loess Plateau, North China, consists of the ancestral temple, the residence of the owner of Jingu Stockade, average residences, workshops, warehouses, etc.

The residence of the owner of Jingu Stockade is composed of four juxtapsed four-in-one courtyards and a wing cave-courtyard made out of the cliff. Only the first courtyard has a main hall while the others all take the cliff caves laid with bricks as their main rooms respectively. The rooms in the first yard are well-proportioned and have symmetrical central axis. The second, third and the fourth yards all have a second gate with festoon curl canopies. In front of the second gate there is a four-meter wide east-to-west sidewalk connecting all the yards, and in the backyard a two-meter wide passageway runs from east to west of the yards. In this way, a huge residence with all the yards independent as well as mutually interlinked with one another is formed.

The Kang Baiwan's Manor is beautiful, imposing and sumptuous. Its exquisite and unique post and panel structure, straight and impressive-looking overlying-ridge gable walls, dainty and exquisite patterns on the doors and windows, refined and gorgeously carved beams and painted rafters, and the elegant and unsophisticated furniture, all add strong local characteristics to the manor.

前院 The front yard

院落花厅 The parlor of the courtyard

内院 The interior yard

民居侧门 The side gate of the vernacular houses

内院宅居 The interior yard of the residence

门楼脊饰 The ridge ornament of the gate-building

窑洞民居 The cave dwellings

围院高墙 The tall wall of the enclosed courtyard

靠山崖窑 The cliff caves lying against slope

侧门 The side gate

河南地坑窑民居
The Underground Cave Dwellings, Henan Province

地坑院是三门峡、巩义一带窑洞民居的主要类型，其中，以陕县南塬庙上村地坑院最为典型，庙上村隶属于陕县西张村镇。

庙上村以张姓人氏为主，地坑院窑洞民居具有悠久的发展历史，百年以上的窑院就有数十座，该村的地坑院传统民居无论是建筑形制、建造质量，还是建筑文化，都是南塬地坑院民居的优秀代表。

庙上村地坑窑的地坑既有正方形，也有长方形的院落，但多数是正方形，于地坑院落四壁开挖8孔或10孔，最多有12孔的窑洞。

The underground cave dwellings are typical houses in the area of Sanmenxia City and Gongyi City in Henan Province, among which the most representative one is the Silo-Yard at Miaoshang Village in the south plateau of Shanxian County. The village belongs to Xizhangcun Town of Shanxian County.

Most people's family name at Miaoshang Village is Zhang. The underground cave dwellings have a long history in this village and several dozens of the houses are more than 100 years old. Whether in the aspect of shape and structure, quality or in the aspect of architectural culture, the traditional vernacular houses in this village are all outstanding models of the underground cave dwellings on the south plateau.

Most of the silo yards in Miaoshang Village shape square, though some of them are rectangular. The four walls in a silo yard are dug to built 8 or 10 underground cave dwellings, some silo yards can have as many as 12 underground cave dwellings.

巩义地坑窑 The underground cave dwellings in Gongyi

庙上村地坑窑 The underground cave dwellings in Miaoshang Village

庙上村地坑窑
The underground cave dwellings in Miaoshang Village

地坑窑入口
The entrance of the underground cave dwellings

窑洞室内布置
The interior arrangement of the cave dwellings

巩义地坑窑
The underground cave dwellings in Gongyi

园林民居

Garden houses

汉族中有一种特殊类型的民居，供世家、文人、士大夫所用，到后代，在一般士民中也有采用，这就是住宅、书斋、庭园结合在一起的民居建筑，称为园林民居，或称庭园民居。它通常又分为三类：

一是宅旁设书斋，也有单独设立者，自成一种类型，称为独立式书斋。它的平面特征是在三开间民居中，将厅堂向前延伸为长厅，在长厅凸出部分左右两旁各设一个小天井，天井内种植纤细的竹木或堆置少量奇石异草，使书斋左右都有庭院绿化。长厅两边都开有较大的窗户，这种书斋虽小，但环境清静幽雅。

二是宅旁设园林，园林独立设置，自成一种类型，它在江南、岭南较多见。其布局特征是园内不设宅居，而以观赏为主，宅与园连接，有门相通，但各自独立。

三是住宅、书斋、庭园组合在一起，三者功能既有联系，又各自独立。它的平面特征是以住宅为主，书斋、庭园为辅。宅内有宁静的环境，书斋内可进行词诗书画朗读品赏，闲时则漫步庭园。

园林的设计，是以厅馆为中心，四周辅以山石、池水、花木、草卉，并用廊、墙、桥、亭为间隔或相连组成各个景区。它要求在占地不大的有限面积内，充分利用周围的环境、建筑和自然景色，并采用组景、对景、借景等手法，创造出更多、更丰富的景色来满足园林主人观赏、可居、可游的目的。

园林民居着重动静结合，以幽雅为主。明代计成的《园冶》一书指出“三分匠、七分主人”，说明园主是创作园林的主导者，又指出“巧于因借，精在体宜”，这是园林创作设计的准则。通过园林民居的营建实践，不但可以获得宜居宜游的艺术享受，也可获得园林民居中的美学品赏价值。

A special kind of houses accommodating thc Han people used to be inhabited by old and well-known families, literati and officialdom (in feudal China). Later, some ordinary citizens also lived in such houses. They are vernacular architectures called garden houses or garden court houses that combine residences, studies and garden courts together. Such houses are classified into three types:

First, a study is built beside a residence and some detached studies have styles of their own as they are built independently. The plane feature of such a study is: in a three-bay house, the hall is extended forward into a long one. On both sides of the jut of the long hall, each small yard is arranged respectively. In the small yards, some slim bamboos, trees and grass are planted and a small amount of stones with odd shapes is banked up so that the courtyard is afforested on both sides of the study. Big windows are installed in both sides of the long hall. Such a study, though small, has a very quiet and peaceful surrounding.

Second, an independent garden with a style of its own, is built beside a residence, which is commonly seen in the region south of the Yangtze River and south China. The feature of its layout is that the garden is mainly ornamental, without residence in it. The residence adjoins the garden opening onto each other with doors, but both the residence and the garden are independent.

Third, the residence, the study and the garden are grouped into one homestead, in which each one is independent with its own distinct functions. It has such a plane feature as to take residence as the principal part and the study and the garden as subsidiaries. The environment within the homestead is quiet and one can read poems, verses aloud and savor painting and calligraphy in the study and stroll idly in the garden at his leisure.

When designing the garden, the hall is taken as the center, complemented with rockery, pond water, flowering wood, grass all around, so that each scenic spot will be formed by spacing the garden with a corridor, a wall, a bridge or a pavilion. It is required that on a limited piece of land, full use be made of the surrounding environment, the architectures and the natural landscape, and the means of the assembled view, the scenic focal point and the borrowed view be adopted so as to create more and richer sceneries to satisfy the garden owners' needs to enjoy the sight of the garden, to live in the garden and to hold parties or games in the garden.

The garden residence emphasizes the association of activity and inertia, focusing quietness and tastefulness. The book named Yuanye by Ji Cheng in the Ming dynasty pointed out that in the creative design of a garden, 30% belongs to the craftsmen while 70% belongs to thc owner, which indicates that thc owner plays the leading role in the design. The book also pointed out that the art lies in cleaver borrowing and the essence rests with suitability, which is the criterion for creating the design of the garden residence. Through the construction practice of the garden residence, one can not only obtain the artistic enjoyment of residence and play, but also savor the aesthetic value of it.

北京恭王府萃锦园

The Cuijin Garden in the Prince Gong's Mansion, Beijing

恭王府始建于清乾隆年间，原为大学士和珅的私宅，建筑布局规整、工艺精良、楼阁交错，充分体现了辉煌富贵的风范。恭王府是我国保存最为完整的王府建筑群，前部为府邸，后部为萃锦花园。

府邸建筑分东、中、西三路，每路由南向北都是由中轴贯穿的多进四合院落组成。中路主要建筑是银安殿和嘉乐堂；东路为多福轩、乐道堂；西路的院落较小，为葆光室和锡晋斋。府邸最深处有一座两层的后罩横楼，东西长达156米。

萃锦园也分东、中、西三路。中路以一座西洋建筑风格的汉白玉拱形门为入口，以康熙御书"福"字碑为中心，前有独乐峰、蝠池，后有绿天小隐、蝠厅，布局令人回味无穷。东路的大戏楼，厅内装饰秀丽，戏楼南端另有园中之园。

The Prince Gong's Mansion was built during Emperor Qianlong's reign in the Qing Dynasty, and served as the private residence of He Shen, one of the grand secretaries at first. As the mansion house of a prince in the Qing Dynasty, the layout is standardized with excellent techniques. With interlocked buildings and open halls, the mansion fully reflects the royal demeanor of splendor, wealth and rank as well as the graceful bearing of elegant style of the people. The Prince Gong's Mansion is the very prince mansion complex that has been preserved most intact in our country, the front part is the impressive-looking mansion house, and the rear part is the deep, quiet and beautiful garden.

The mansion is divided into three layouts, east, middle and west, each having several rows of four-in-one courtyards through which the axile runs strictly from south to north. The principal architecture in the middle zone are the Yin'an Palace and the Jiale Hall. The front yard in the east zone is the Duofu Corridor, and the rear yard is the Ledao Hall, which used to be the living room of Yixin, the Prince Gong. The four-in-one yard in the west zone is relatively small and exquisite whose main architectures are the Baoguang Room and the Xijin Room. In the deepest place stand a two-story posterior block with a length of 156 meters from east to west, 88 windows built in the back wall. In the posterior block, there are 108 room, but is commonly called 99 and half rooms, with the meaning of "brimming over when something is full" from the Taoism.

Echoing each other with the mansion house, the Cuijin Garden in the Prince Gong's Mansion is also divided into three layouts. The middle zone takes the white-marble arched stone gate of a building with western style as its entrance, the stone tablet with the character of "福" written by Emperor Kangxi in the Qing Dynasty himself as the center, with the Dule Peak and the Fu Pool in front of the building, the green concealment and the Fu Hall at the back of it so that the layout leads people to endless aftertastes. The interior of the grand theater in the east zone is decorated delicately and beautifully, and on the southern end of the grand theater, the five beautiful views of the Mingdao Room, the winding path leading to secluded spots, the pleasing trees by the boulevard, the Yinxiangzuiyue and the Liubei Pavilion form a garden within a garden.

恭王府大门 The gate of the Prince Gong's Mansion

宅居大门 The gate of the residence

戏楼小院　The theater in the small courtyard

园林檐廊 The eave veranda in the garden

内院宅居　The interior yard of the residence

内院宅居　The interior yard of the residence

园林戏楼　The theater in the garden

内院垂花门　The floral-pendant of the interior yard

宅居侧门　The side gate of the residence

园林假山　The artificial hill in the garden

山东济南万竹园
The Wanzhu Garden, Jinan, Shandong Province

万竹园位于济南市西青龙街，由前院、东院、西院和花园组成，始建于元代，经修复改建后纳入趵突泉公园内。

万竹园为山东庭院式住宅的典型，它汲取中国园林“小中见大”的造园技法，巧妙布置楼、堂、亭、榭、廊、桥等不同的建筑形式，把宅园划分成具有不同功能的三合院和四合院，院院相通，溪流相连。空间灵活多变，大小高低，明暗开合，风格各异。泉水穿庭入户，环绕于庭院之中。亭榭则被巧妙地镶嵌在建筑群内，人居亭中，驻足赏鱼，别具情趣。水体形态多变，有动有静，运用自如。植物则以竹为基调。

花园居西部，空间开朗，松竹掩映。东南溪流围绕，西北则游廊环抱，中间地形起伏。绿草如茵，花间隐榭，怪石嶙峋，景色宜人。

The Wanzhu Garden is situated in the West Qinglong Street of Jinan, consisting of the front yard, the east yard, the west yard and the garden. Its construction began during the Yuan Dynasty and later it was renovated and transformed as a part of the Baotuquan Garden.

The Garden of Ten Thousand Bamboos is the typical example of the garden houses in Shandong, which draw the Chinese skills of making gardens “much in little”, and skillfully arrange different architectural styles such as terraces, open halls, pavilions, waterside pavilions, corridors, bridges, etc. to divide the residence into three-in-one and four-in-one courtyards with different functions. The yards are connected and the streams join each other. Flexible are the spaces of the yards that vary in styles, sizes, heights, brightness and openness. Spring water passes through gardens into homes or embraces courtyards and pavilions are skillfully embedded among the houses. When one stands in a pavilion, and appreciates schools of fishes in the water, he will have a unique temperament and delight. The water body varies in patterns, sometimes ripples and riffles, sometimes remains calm and tranquil. Plants are basically bamboos in the garden.

The spacious garden is situated in the west, pines and bamboos setting off one another in it. In the garden, streams embrace the southeast, the covered corridors encircle the northwest and the landform in the middle rises and falls, with a carpet of green grass, dimly water-pavilions among flowers, jagged rocks of grotesque shapes and attractive scenery.

万竹园水庭 The river courtyard of the Wanzhu Garden

万竹园内庭 The interior courtyard of the Wanzhu Garden

水院石桥　The stone arch bridge of the water garden

民居窗花装饰　The window decorations of the vernacular houses

水庭凉亭细部　The details of the pavilion of the river courtyard

水院大门　The gate of the water garden

院墙砖雕　The brick carvings of the garden walls

宅居内院　The interior yard of the residence

宅居二门　The second gate of the residence

水庭凉亭　The pavilion of the river courtyard

护院渠水　Water in the ditch of the protective yard

万竹园院门 The gate of the Wanzhu Garden

江苏扬州个园
The Geyuan Garden, Yangzhou, Jiangsu Province

江苏扬州个园是清嘉庆二十三年（1818年）大盐商黄应泰在明代寿芝园旧址上兴建起来的。园主爱竹，园内遍植翠竹，意取东坡诗句："宁可食无肉，不可居无竹。"因竹叶形状像"个"字，故名"个园"。总体布局，宅居在前，园林在后。宅居分东、西两路，通过中间的巷道通往后部园林。

个园面积约30亩，由于布局巧妙，显得曲折幽深，引人入胜。扬州园林素以叠石为胜。个园以四季假山的堆叠精巧而著名。造园工匠们选用褐黄石、太湖石、雪石和状如竹笋的石笋，叠成四组假山，表现春夏秋冬四季景色，称为四季假山。

春景，在竹丛中选用石笋插于其间，取雨后春笋之意。夏景，在浓荫环抱的荷花池畔，叠以湖石，使人感到仲夏的气息。秋景，黄褐山石峰峦起伏，登山眺望使人有秋高气爽之感。冬景，用白色的雪石堆砌，远看犹如留在石上的残雪。绕园一周如经历了春夏秋冬四季。

The Geyuan Garden in Yangzhou, Jiangsu Province was built in the 23rd year of the Jiaqing Empero's reign of the Qing Dynasty (1818) by Huang Yingtai, a salt merchant on the base of the Shouzhiyuan Garden of the Ming Dynasty, in which green bamboos grow everywhere, and the implication comes from the poem by Su Dongpo "better to have my meals without meat than have my garden without bamboo." The garden is named as "个园 (Geyuan)" only because the bamboo leaves resemble the character of "个". The layout is that the front part is the residence while the back part is the garden. The residence is divided into two layouts: the east and the west. The ally in the middle leads to the garden in the back part.

The area of the Geyuan Garden is 30 mu (20000 m^2) but looks intricate, deep, quiet and bewitching thanks to the ingenious overall arrangement of it. The gardens in Yangzhou have long been known for their up-piled rocks and the Geyuan Darden is famous for the ingeniously up-piled four-season rockeries. The garden's craftsmen selected some isabelline rocks, lake rocks, snow rocks and stalagmites resembling the bamboo shoots and form four groups of rockeries by piling them up, representing the scenes in spring, summer, autumn, and winter. That's why they are called four-season rockeries.

Stalagmites were selected and put in the bamboo groves to form the spring scenery with the implication of bamboo shoots after a spring rain. At the lotus pond surrounded by thick branches and leaves, some lake rocks were piled up to make people feel the tinges of summer. That's how the summer scenery came into being. Ridges and peaks of the isabelline rocks rise one after another, people can sense autumn scenery of the clear and crisp autumn climate by climbing up and overlooking from the mountains. The winter rockeries were piled up with white snow rocks, seen from a distance, they are just like rocks with some relict snow left on them. Walking around the garden, one may feel as if he had experienced four different seasons in a whole year.

清漪亭 The Qingyi Pavilion

园林入口春景图 The spring landscape picture of the entrance of the garden

园林洞门 The artistic door of the garden

丛书楼 The Congshu Building

江苏苏州
网师园
The Wangshiyuan Garden, Suzhou, Jiangsu Province

网师园位于苏州市十全街，于南宋时期始建，称"渔隐"，清乾隆年间重建，改名"网师园"。园在住宅西面，以布局紧凑、建筑精巧和空间尺度比例良好著称，是当地中型园林的代表。

园中部凿池，岸周叠砌石矶、假山。其东南与西北有水湾，曲折深奥。池东为原住宅厅堂，有大门、轿厅、大厅、花厅及庭院两重。花厅西北侧有小门可通园内。池南为"小山丛桂轩"，是园中主厅，四周罗石植桂，环境甚为清幽。循向北，至"濯缨水阁"，此阁建于湖上，凭栏可观水中游鳞及环池景物。该园从各个角度都能构成良好画面，这是它的设计有着良好、周密的构思所致。

西北小院过去以盛植芍药闻名，现置峰石、花台及半亭。北端之殿春簃乃主人读书之所，深庭静逸，竹石清幽。

The Wangshiyuan Garden is situated in the Shiquan Street of Suzhou. Its construction began during the Southern Dynasty and called Yuyin Garden. It was rebuilt during Emperor Qianlong's reign and renamed the Wangshiyuan Garden. The garden is on the west of the residence, a representative of the local medium-sized gardens, famous for its compact layout, exquisite construction and good proportion of the space dimension.

In the center of the garden there is a pond, around the bank of which rocks and rockeries are built. A water brace runs deep windingly from the southeast to the northwest. The main hall of the residence with the main entrance, the gate for sedan, the parlor and two courtyards used to be on the southeast of the pond. On the northwest a small exit leads to the yard. On the south of the pond stands Xiaoshan Conggui Xuan, the main hall with rocks and sweet osmanthus trees all around, which has a very quiet environment. Going north to the "Guanying Water Terrace", which is built on the water, one can enjoy the sight of the swimming fish and the views around the pond by leaning against the railing of it. From all angles, the garden can form a fine picture as the result of fine and thorough conception about the successful design.

The small courtyard on the northwest used to be famous for planting Chinese herbaceous peonies, but now rocks, a flower bed and a pavilion are set up there. The Chunyi Palace on the north end, with a deep and quiet environment and beautiful bamboos, is the place where the owner of the residence used to study.

园林水景 The water scene of the garden

庭院轩馆 The pavilions in the garden

庭园路径　The routes in the garden

门楼砖雕　The brick carvings on the gate-building

花墙连廊　The tracery veranda

江苏吴江
震泽师俭堂
The Shijian Manor of Zhenze Town, Wujiang County, Jiangsu Province

师俭堂位于江苏吴江震泽镇老街宝塔街上，宅居主人姓徐，是震泽的望族。师俭堂坐北朝南，三面临河，可前门上轿，后门下船，水陆称便。集河埠、行栈、商铺、街道、厅堂、内宅、花园、下房于一体的建筑群组，反映了晚清工商绅士坐行经商的特点。

师俭堂布局规整严谨，占地2700余平方米，共有大小房屋147间。师俭堂面阔五间，六进高墙深宅。若每进开敞大门，可从河埠一直望到第六进厅堂上的屏门。师俭堂内部装饰华美，各类雕刻精雅别致、形态生动，人物花鸟，惟妙惟肖。

宅内花园取名锄经园，小巧玲珑。亭台楼阁，游廊假山，花卉树木，一应俱全。假山下有山洞，上设半亭，高低起伏，错落有致，闲庭信步，别有情趣。

The Shijian Manor is situated at a historic street named the Pagoda Street in Zhenze Town, Wujiang County, Jiangsu Province. The surname of the owner of the manor is Xu, the respected and influential clan in Zhenze. The manor faces south, with a river surrounding it on three sides. The transportation in the manor is very convenient, since the owner could get on his sedan in front of the main gate and get off his boat from the back gate. This architectural complex incorporates the wharf, broker's storehouses, stores and shops, streets, halls, inner chambers, gardens and the servants' lodgings, reflecting how the businessmen and gentlemen engaged in trade during the late period of the Qing Dynasty.

With an area of over 2700 m^2, and a total number of 147 rooms, the layout of the Shijian Manor is standardized and rigorous. The manor is a six-row compound with high walls, and the front of the main hall consists of five bays. If all the gates are open, the screen gate in the hall of the sixth row can be seen from the wharf. The interior of the manor is elegantly decorated with various delicate carvings, vivid forms of absolutely lifelike figures, flowers and birds.

The garden within the manor is named Chujing Garden, little and dainty. Pavilions, terraces and open halls, verandas and rockeries, flowers and trees, are all available in all varieties. There is a cave under a rockery, with a semi-pavilion constructed on it. With ups and downs, these sceneries are well-proportioned with different heights. Strolling idly in the courtyard, one will find the manor has a particular flavor.

前院门楼　The gate-building of the front yard

庭园曲廊　The wavy corridor in the garden

后院洞门　The artistic door in the back courtyard

庭园花厅　The flower hall

内院天井　The small yard in the interior yard

宅居偏厅　The wing hall of the residence

师俭堂后院　The back courtyard of the Shijian Manor

庭园院门　The gate of the garden

浙江杭州
胡雪岩故居
Hu Xueyan's Former Residence, Hangzhou, Zhejiang Province

胡雪岩，杭州人，祖籍安徽绩溪，少年时到杭州当伙计，因勤奋经营，遂成为富商。其故居是住家、庭园合一的建筑群体。

胡雪岩故居建于清同治十一年（1872年），分为三部分，主轴线由轿厅、照厅、百狮楼（正厅）、东西四面厅所组成，其东部为住家生活用地，西部为花园。主体建筑继承了中国传统建筑的形制和做法，也借鉴了外来建筑文化、技术和材料。宅内装饰、装修包含木雕、砖雕、石雕、灰塑、彩画，用材高档，工艺精致，图案和谐，色彩艳丽。

西部花园以山水景象为主题，使水面与空间相互渗透，似分似连。园中有假山，做成人工溶洞，依山筑楼阁，临水建轩厅，山水与建筑既有对比，又有节奏，整座园林建筑与地方特色相互交融与和谐。

Hu Xueyan, a citizen of Hangzhou, his ancestral home was Jixi County, Anhui Province. When he was a teenager, he came to Hangzhou and began working as a shop assistant. Through diligence and frugality, he started his own business and became a rich merchant. His residence is a complex combining the household with a garden courtyard.

Built in the 11th year of Emperor Tongzhi's reign in the Qing Dynasty (1872), Hu Xueyan's Former Residence consists of three parts: On the axis stand the Sedan Hall, the Zhao Hall, the One-hundred Lions Building (the main building). The east part is the residence while the west is a garden. The main buildings carried forward the shape and structure of the traditional Chinese architecture and simultaneously absorbed the foreign architectural cultures, skills and materials. Superior-quality materials were applied in the interior decoration and furnishing, including the wood carving, brick carving, stone carving, clay sculpture, and the color painting, and therefore it was done with superb craftsmanship, harmonious patterns and gorgeous colors.

The mountain-and-water scene is the main theme of the garden in the west part, in which the water surface and space permeate each other, as if they were not only divided but also connected. The rockeries in the garden have an artificial karst cave. The pavilions, terraces and open halls are built against the mountains or by the water. The mountains and water not only form a sharp contrast but also have a lively rhythm. The architectural tradition of the whole residence blends with the local feature harmoniously.

西园水景 The water scene of the west courtyard

宅居西园　The west courtyard of the residence

西园楼阁　The attic in the west courtyard

门厅入口　The entrance of the door hall

门厅天井　The small yard of the door hall

宅居外檐　The outer eaves of the residence

花厅檐廊　The eave veranda of the flower hall

天井　The small yard

厅堂　The hall

宅居内院　The interior yard in the residence

照壁砖雕　The brick carvings on the screen wall

厅堂后院　The back yard of the hall

厅堂后院　The back yard of the hall

浙江湖州
南浔小莲庄
Xiaolian Village of Nanxun Town, Huzhou City, Zhejiang Province

小莲庄位于浙江省湖州市南浔镇镇西南古桥西，建于1885～1924年间。小莲庄由园林、家庙和义庄三部分组成。

园林部分有外园和内园，外园以荷花池为中心，池水清碧，满植藕莲，亭、廊、房、楼绕池布置，石径弯曲，池面宽广，建筑朴实，疏朗有序。西岸有长廊，东有石桥小榭。内园位于东南角，以假山为中心，北有高墙与外园相隔，假山玲珑峭削，山道盘旋，山顶有小亭，可远眺田野。

花园西部为刘氏家庙，门前有两座石雕牌坊，精致小巧，虽建造在狭窄的墙门通道内，但显得高耸肃穆。家庙共三进，后厅为悬挂祖宗画像之所。

家庙西为义庄，共两进，再往西为后园，河渠清澈，古樟高大，清净幽深。

Xiaolian Village, built during the years from 1885 to 1924, is located on the west of the historic bridge in Nanxun Town, Huzhou City, Zhejiang Province. The village consists of three parts: the garden, the ancestral temple and Yizhuang Village.

The garden is subdivided into the outer garden and the inner garden, the former takes the lotus pond as its center, in which the water in the huge pond is clear and clean, packed with lotuses; the pavilions, veranda, houses and winding stone paths are arranged around the pond with simple patterns and reasonable spacing. A gallery is on the west bank of the pond and a stone bridge with a pavilion on the east bank. The latter is located in the east corner, taking the rockery as its center, with high walls screening the outer garden on the north. The rockery is ingeniously and delicately wrought, and rises steeply, with mountain passes circling around, and a little pavilion standing on the top of the mountain from which one can overlook the fields far away.

On the west part is the Liu clan's ancestral hall, in front of which stand two stone-carved memorial archways, little and dainty, towering, solemn and respectful though built within the narrow passage to the wall gate. The ancestral hall has three rows and the back hall is the place where the ancestors' portraits are hung.

Situated west to the Liu clan's ancestral hall is Yizhuang Village, consisting of two rows. Further west is the clean, deep and quiet back garden, in which the ditch water is limpid and the camphorwood trees are tall.

小莲庄荷花池 The lotus pond of the Xiaolian Village

大门入口 The entrance of the gate

傍水亭榭　The pavilion beside the stream

花厅藻井 The patio in the flower hall

花厅陈设　The furnish of the flower hall

园林曲径　The winding paths in the garden

园林连廊　The veranda in the garden

家庙牌坊　The memorial gateway of the ancestral temple

园林连廊　The veranda in the garden

云南建水朱家花园

The Zhujia Garden, Jianshui County, Yunnan Province

建水古称临安，元代以来就是滇南政治、军事、经济、文化和交通中心，这里有元、明、清各代古建筑近百处，古桥50余座。

朱家花园，规模宏大、布局合理、设计精巧，是具有江南园林特色的古建筑群。它坐落在建水县城建新街中段，清光绪年间所建。房舍鳞次栉比，建筑屋角起翘，陡脊飞檐，雕梁画栋。园内建筑呈"纵三横四"布置，院落之间由巷道所连接，让人感觉大院深不可测。

家宅东面为宗祠，三进院落，前有一方小水池，称"小鹅湖"，小鹅湖旁建有水榭，水榭实际上是一座玲珑小巧的水上戏台，台口呈八字形，由四根石柱架在水中，柱头上饰以石象、斗栱和插梁，雕琢精细，实为民居建筑中不可多得的精品。

Jianshui used to be called Linan in the ancient times and has been the center of politics, military, economy and transportation of south Yunnan since the Yuan Dynasty. There are nearly 100 historic buildings and over 50 historic bridges built in the Yuan, Ming and Qing Dynasties respectively.

The large-scaled, reasonably-designed and well-planned Zhu's Garden is a complex of historic buildings with the characteristics of the gardens south of the Yangtze River, situated in the middle section of the Jianxin Street of the county town of Jianshui County, built during the reign of Emperor Guangxu in the Qing Dynasty. The houses are arranged in close order. The corners of the houses are raised, the ridges are steep and the eaves are upturned with carved beams and and painted rafters. The houses in the garden are arranged in three-vertical and four-horizontal rows. The yards are linked with alleys, giving a fathomless feeling.

On the east side is the ancestral hall which is a three-row courtyard, with a small pond called the Little Goose Lake in front of it. By the pond a waterside pavilion is built, which in fact is an exquisite stage on water whose front is splay erected with 4 stone pillars. The heads of the pillars are decorated with elegant engraving of stone elephants, bucket arches and inserted beams, which is really rare cream of vernacular architecture.

入口大门　The entrance of the gate

朱家花园“梅馆”　The “Mei Hall” of the Zhujia Garden

戏台　The theater

大门入口照壁小院
The screen wall of the small courtyard at the entrance of the gate

家祠敞厅
The open hall of the ancestral hall

宗祠前“小鹅湖”水池　The pond called the Little Goose Lake in front of the ancestral hall

宗祠厅堂　The hall of the ancestral hall

宅院“竹园”　The "Bamboo Garden" of the courtyard

入口大门檐廊　The eave veranda at the entrance of the gate

广东番禺
余荫山房
The Yuyin Shanfang Villa of Panyu District, Guangzhou, Guangdong Province

余荫山房位于广东番禺南村，建于清同治年间。

进入余荫山房正门，过厅堂，穿竹径，到达山房的花园门。门旁有对联一副："余地三弓红雨足，荫天一角绿云深"，可说是对园中主景的写照。

山房园地仅2000平方米，但亭桥楼榭，曲径回栏，荷池石山，名花异草，一应俱全。园景分东、西两部分，西半部以长方形石砌荷池为中心，池南有造型简洁的临池别馆，池北为主厅深柳堂。深柳堂前庭有两棵苍劲的炮仗花古藤，花儿怒放时宛若一片红雨，十分灿烂绚丽。东半部的中央为一八角形水池，池中有八角亭一座，名"玲珑水榭"。水榭东南沿园墙布置一假山，水榭东北有来薰亭和孔雀亭，周围还有多株大古树。东西两半部的景物和水面则有"浣红跨绿"拱廊小桥连接，在园中起到了画龙点睛的作用。

Yuyin Shanfang Villa is situated in Yunan Village of Panyu District in Guangzhou, Guangdong Province, built during Emperor Tongzhi's reign in the Qing Dynasty.

Entering the main entrance of Yuyin Shanfang Villa, passing through the hall and the path lined with bamboos, one can get to the gate of the villa's garden. There is a couplet that means "on the earth, the three gong (an old unit of length for measuring land, equal to about 1.67 meter) of land is sufficient with blood-rain; the corner of the sky is deep with green clouds", which can be said as a portrayal of the main scenery of the garden.

The villa only takes up an area of 2000 square meters but pavilions, terraces, open halls, winding paths, ambulatories, the lotus pond, rockeries, well-known flowers and extraordinary grasses are available in all varieties. The garden scenery is divided into two parts: the west and the east parts. The former part focuses itself on the lotus pond laid with rectangular slabstone, with the concise-modeling Linbie Hall close to the pond on the south and the Shenliu Hall (the main hall) on the north. In front of the Shenliu Hall grow two old and strong cracker-flower vines, which are splendid and magnificent, just like a scene of blood-rain when the flowers are in full bloom. In the center of the latter part is a octagonal pond, in which stands an octagonal pavilion named "Linglong Water Pavilion". A rockery is arranged along the garden wall on the southeast of the water pavilion, on the northeast of which stand the Laixun Pavilion and the Peacock Pavilion surrounded with many old trees. The sceneries and the water surface of both parts are linked with the Huan-hong-kua-lu arch bridge (the name of the bridge means the garden boasts itself for its beautiful scenery with colors of red and green), which plays an important role of adding the finishing touch in the garden.

玲珑水榭八角池　The octagonal pond called "Linglong Water Pavilion"

浣红跨绿廊桥　The gallery bridge called Huan-hong-kua-lu

深柳堂客厅　The drawing room of the Shenliu Hall

玲珑水榭室内　The interior of the Linglong Water Pavilion

园林连廊　The veranda of the garden

台湾台北林家花园

The Lin Clan Garden in Taipei, Taiwan Province

台北林家花园是台湾林本源家族大厝的后花园，清光绪十九年（1893年）建成，是台湾最有代表性的庭园。

进入庭园，第一景是用于藏书的汲古书屋，此后经游廊向左到方鉴斋，斋前有莲花池，池中设戏台。沿游廊继续前行至来青阁。来青阁是园中最精致的建筑，为二层阁，楼上全部以樟木、楠木建造，华丽非凡。过来青阁，回廊分为两路，一路向北经香玉簃、定静堂至月波水榭，一路向西，跨陆桥，经观稼楼到定静堂及大池。

香玉簃为观赏奇花异卉之处。月波水榭在定静堂东侧，是坐落于水池中的双菱形建筑，有桥可通。定静堂为四合院建筑，是林家宴客的场所。观稼楼较来青阁小，前有轩亭，后有庭院。庭院的终点为榕荫大池。池为不规则形，中有小岛及半月桥，设有码头，池中可泛舟。庭园的后门在定静堂北侧，门额题曰“板桥别墅”。

Built in the 19th year of Emperor Guangxu's reign in the Qing Dynasty (1893), the Lin Clan Garden, in Taipei, the most typical courtyard of Taiwan, is the back garden of Lin Benyuan's clan residence.

The first scenery in the garden is the Jigu Study that is used for collecting books. Go left via the veranda and one can get to the Jian House. In front of the Jian House there is a lotus pond with a theater built over a part of it. Go straight further via the veranda and one can get to the Laiqingge Builing, the most elegant architecture in the garden. It has two floors and the second floor, built with camphorwood and nanmu, is extremely resplendent. The veranda is divided into two routes after the Laiqingge Building. One route goes north to the Yuebo Waterside Pavilion via the Xiangyuyi (a small connecting room between top apartments) and the Dingjing Hall. The other leads you west to the Dingjing Hall and the big pond via the Guanjia Building after crossing the Lu Bridge.

The Xiangyuyi Room is where one can enjoy the sight of rare flowers and extraordinary grasses. To the east of the Dingjing Hall stands the Yuebo Water Pavilion, a double-diamond-like architecture situated in the pond, with a bridge leading it to the bank. The Dingjing Hall is a four-in-one courtyard architecture, used for the Lin's family to entertain their guests. The Guanjia Building is smaller than the Laiqingge Building. In front of the Guanjia Building there is a small pavilion and behind it is the courtyard, the end of which is the Rongyin Big Pond. The form of the pond is irregular, in which there is a small isle and a half-moon-shaped bridge. There is also a ferry and people can sail off in a boat in the pond. The back gate is to the north side of the Dingjing Hall, on whose horizontal tablet four Chinese characters “板桥别墅 (Mr. Zheng Banqiao's Villa)” were inscribed.

园林亭阁 The pavilion in the garden

林家花园后园水景 The water scene of the back yard of the Lin Clan Garden

园林门洞　The gateway of the garden

月波水榭　The Yuebo waterside pavilion

定静堂　The Dingjing Hall

防御民居

Defensive vernacular houses

汉族先民在南迁过程中，为求生存发展而在偏远山区或沿海人口稀少地区聚集而居，形成了一种特殊类型的民居即防御性民居。这种民居最大的特征是聚居和防御。其布局是：在山地，设在高处，在平地，为多层高屋。其外形特征：外围有高大深厚的围墙，单独的出入大门。一、二层不设窗户，三、四层开八字形小窗供瞭望用。平面较大，内有大院，可供多户聚居，甚至数十户使用。大院内设有祖堂，供祀祖先。在北方称为堡寨，南方称为围楼。

北方堡寨大多为靠山面水，利用地形，有别于平地大院的一种围合型建筑，防御性很强，上有碉堡，四周有望楼，或设置用于瞭望守备的高楼亭阁，周围有高厚的围墙。出入只有一个大门，有的在大门外还挖有较深的壕沟。户主多为豪绅官宦，有的称城堡，有的称庄园，在四乡宗族村民聚居之地统称为寨，因人口众多，寨内房屋毗连，并有小街小巷。

堡寨外观，壁垒森严，虽有四角望楼阁亭点缀，但仍呈现出威赫之势。在山区周围之民间堡寨，虽有封闭围墙，但外观呈现出一种厚实、淳朴、粗犷的面貌。

南方防御性民居以客家围楼为代表，分布在粤、闽、赣三省交界地区，大部分在山区。福建西南地区称之为土楼，有圆形、方形、椭圆形、六角形等，有单个存在的，有两个甚至多个组合毗邻在一起的，最多的如福建永定县下洋镇初溪村楼群，有多个围楼排列在一条山沟里，远望十分壮观。江西南部地区称之为围子，以长方形围楼为主，一般比较大。广东东北部梅州地区称之为围龙屋，正屋为三进院落式建筑，两侧加横屋，有单列，有双列，前有地坪，半圆形池塘，后有半圆形屋，背靠大山，以林木作围蔽。远望这些民间围楼，体形高大，但土坯墙体、灰瓦屋面，也有刷白墙面，总的来说，外观反映出了民间一种淳朴、厚实、粗犷的自然之美。

In order to survive in the harsh and cruel environment, in the southward migration, the ancestors of Han people usually settled collectively in the remote mountainous areas or sparsely populated coastal areas. Thus it developed a special type of dwelling, the defensive vernacular house to serve the purpose of cluster living and defense. As to its layout, it is built either on higher place in mountain area or as high-rise on the flat land. The house is characterized by tall and thick external walls and independent entrance door. There is no window on the first and second stories and only on the third and fourth stories, there can be found small 8-shaped windows installed for lookout. Thanks to its relatively large internal compound, several and even dozens of families can live together and share one courtyard. Besides, the ancestral hall is also built inside the compound. Such a vernacular house is called Baozhai (village with surrounding wall) in north China and known as Enclosed House in south China.

Different from the enclosed compound on the flat ground, Baozhai in the north, taking advantage of the terrain, is mostly constructed against the mountain and facing the water, having a strong defensive function. In terms of appearance, it has blockhouse and belvedere in four directions, or high-rise pavilions as lookouts, surrounded by high and thick walls. As there is only one entrance door, a deep groove is dug right outside the main entrance in front of some houses. These houses used to be occupied by rich people or imperial officials, named as castle or manor. In the places where the villagers from the same clan inhabit collectively, it is known as Zhai (the village). Because of the large population, the houses inside Zhai are jointed with each other, forming small streets and lanes.

The closely guarded Baozhai, although decorated with belvedere in four corners, appears imposing and glorious. However, Baozhai in the mountainous area, encircled by walls as well, presents a different appearance, looking solid, simple and rugged.

The defensive vernacular house in the south is well represented by Hakka Enclosed House, which is mostly distributed in the mountainous border areas of Guangdong, Fujian and Jiangxi provinces. Known as Earth Building in the southwest region of Fujian and in the shape of round, square, oval, hexagonal, etc., it is constructed either independently or in groups of two or more. The largest cluster of these buildings is found in the Cuxi Village, Xiayang Town, Yongding County of Fujian. A number of enclosed houses are arranged in a ravine, making a spectacular view. Weizi is the name given to it in the south of Jiangxi, which is mainly in rectangular shape and large in size. In the Meizhou area of northeast of Guangdong, it is known as the Weilong Building. Its main building is a three-layered courtyard architecture, with sidelong rooms on both sides in a single row or double rows, flat ground and semicircular-shaped pond in the front and semicircular ridge house in the back, lying against the mountain and being encompassed by woods. From afar, these houses are big in size, with adobe walls and gray tile roofs, or walls painted in white, mirroring a simple, thick, rugged and natural beauty of the countryside.

福建永定五凤楼

Wu Feng Lou (five-phoenixtower), Yongding County, Fujian Province

五凤楼的分布范围较广，主要分布在永定县、南靖县、平和县和诏安县等乡镇，在福建客家腹地上杭、连城等地也有存在。

五凤楼采用中轴线对称的布局手法，以正楼，即后堂为主体建筑，前为门楼、厅堂，两侧为对称的横屋，横屋数量2～6列不等。这种府第式土楼，院落重叠，主次分明，反映出严格的等级意识。

五凤楼的造型在中轴三堂式的基础上发展为多种形态，其建筑立面花样繁多。前堂是门厅，后堂及其左右堂屋为两层，横屋为平房。有的后堂整座为三层，有的后堂的中心一堂为三至四层，而后堂左右则为两层、三层，呈官帽状。整体建筑群前低后高，层层跌落，高大对称，气势雄伟。

Wu Feng Lous are widely distributed, mainly in Yongding County, Nanjing County, Pinghe County and Zhao'an County, as well as in the hinterland of the Hakka area and Liancheng County of Fujian Province.

Wu Feng Lous adopt axial symmetry in the layout. The principal building, or the back hall is the main building and in the front are the gate building and the hall. On both sides are strip-shaped houses parallel to the central axis known as horizontal houses which are left-right symmetrical and stand in two to six lines. Such mansion-style earthen buildings are deployed with courtyards overlapping each other and in a prioritized manner, reflecting strict hierarchy.

The sculpt of the Wu Feng Lou develops from three halls on the central axis into various styles with diversified facades. The antechamber is a door hall. The rear-chamber and main rooms at both sides are of two stories and horizontal houses are of one story. In some Wu Feng Lous, the rear-chamber has three stories. In others, the central hall of the rear chamber has three to four stories while the halls at the two sides have two to three stories, shaped like an official's hat. The entire complex is lower in the front and higher at the back, layer upon layer, being tall and symmetrical, showing an imposing grandeur.

福建永定五凤楼
Wu Feng Lou (five-phoenix tower) in Yongding County, Fujian Province

福建客家土楼

The Hakka Tu Lou (earthen towers), Fujian Province

福建客家土楼主要分布在闽西、闽南一带客家人独居及客家人、闽南人混居地区。

土楼有圆楼、方楼等多种形式。形态特点可归纳为内通廊式和规整平面式。外墙体采用生土夯筑，内部为木构架，层数通常为3～5层。土楼内同族人聚居，每户平均分配房间面积。在中轴线上有供全族进行节庆活动的堂屋、天井和祖堂。

福建土楼一直为客家人喜爱和沿用，因为它具有以下特点：①在内居住舒适方便，并具有通风、采光、抗震、防潮、隔热、御寒等多种功能。②就地取材，施工方便，节约能源，节约用地，利用无污染材料，保护自然环境。同时，经济实用，适应广大农村经济水平。③堡垒式的外观、厚重的墙体保卫人们的安全。④强烈的中轴对称，以厅堂为中心布局，满足中国传统的宗法观念和聚居习俗。

The Hakka Tu Lous are mainly distributed in the western and southern parts of Fujian Province where the Hakka people live alone or live with the Minnan people.

A Tu Lou is built in round, square and other forms. Its characteristics can be summarized as having inner corridors and an orderly layout. Its outer wall-body uses adobe earth and the inner wall-body wooden framework, usually of three to five stories. Family members of the same clan live together in a Tu Lou and each household will be allocated an average living space. On the axial line, there are a main room, a patio and an ancestral hall where the festivals of the clan are celebrated.

The Hakka Tu Lous in Fujian have been adored and used by the Hakka people generation after generation because it has the following characteristics: first, it is very comfortable and convenient for the people to live in with the functions of ventilation, lighting, anti-earthquake, anti-moisture, heat-proof, cold-proof and others. Second, it is very easy to build as the construction materials are local. It is energy-efficient, land-saving, pollution free and environmentally friendly. Meanwhile, it is economical and practical, suitable for the economic level of the vast rural area. Third, its fortress-like appearance and thick walls can protect the inhabitants from intrusions. Forth, axial symmetry and hall-centered layout cater to the traditional Chinese patriarchal values and cluster-settlement custom.

福建南靖河坑村土楼群 The Tu Lou Cluster in Hekeng Village, Nanjing County, Fujian Province

福建南靖塔下村土楼群 The Tu Lou Cluster in Taxia Village, Nanjing County, Fujian Province

福建客家乡村聚落 The village settlement of the Hakka in Fujian Province

福建南靖塔下村祠堂
The Ancestral Temple in Taxia Village, Nanjing County, Fujian Province

永定客家方楼
The square Tu Lou in Hakka Yongding County

永定客家土楼天井院落
The small yard of the Hakka Tu Lou in Yongding County

永定光裕楼 The Guangyu Building in Yongding County

福建永定振成楼

The Zhencheng Building, Yongding County, Fujian Province

振成楼坐落在永定县湖坑乡洪坑村，始建于1912年，历时5年建成。

振成楼的内部空间是一座八卦形的同圆心内外两环的土楼。外环楼四层，共184个房间。内环两层，有32个房间。外环按八卦方位，用砖墙分隔成8个单元，楼中对称地布置4部楼梯。走马廊通过隔墙的拱门连通，青砖隔墙起到了隔火的作用。走马廊的木地板上还加铺一层地砖，也起到了防火作用。第三、四层走马廊的栏杆做成客家土楼中不多见的美人靠。

内环楼由两层的环楼与中轴线上高大的祖堂大厅围合而成，祖堂平面为方形，攒尖屋顶，正面四根立柱采用西洋古典柱式，柱间设瓶式栏杆。内环楼二层的回廊采用精致的铸铁栏杆，这些中西合璧的做法在客家土楼中都是少有的。

The Zhencheng Building is located in Hongkeng Village, Hukeng Town, Yongding County of Fujian Province. Its construction began in 1912 and was finished five years later.

The interior space of the building is centripetal with two rings arranged in an Eight Diagram layout. The outer-ring has four-stories and 184 rooms and the inner ring two stories and 32 rooms. Arranged in the form of an Eight Diagram, the outer ring is separated into eight units by brick walls and four stairs are placed symmetrically. The Zouma corridors in different units are connected with each other through the arched door in brick walls which also serve as fire walls. The floor of the Zouma corridors is exceptionally covered with a layer of brick to prevent fire. The railings on the third and forth floor have been constructed with meiren kao (banisters that win the name for beauties to lean along their waist) which are rare in the Hakka earthen buildings.

The inner ring is enclosed by a two-storey ring-shaped building and the tall ancestral hall on the axial line. The ancestral hall is square-shaped with a pointed roof. Four columns in the front side are western-style classical columns, with bottle-shaped railings setting between them. The corridor on its second floor uses exquisite cast-iron railings. These practices mixing the oriental and western styles are rare among the Hakka Tu Lou.

永定振成楼围屋

The encircled houses of the Zhencheng Building of Yongding County

围屋内檐走马廊

The Paoma corridor of the inner eaves of the encircled houses

福建永定福裕楼
The Fuyu Building, Yongding County, Fujian Province

福裕楼位于福建省永定县湖坑乡洪坑村，建成于1882年。该楼是五凤楼发展到方楼的过渡类型。五凤楼的下堂在这里变成了两层楼房，与两侧三层的横屋相连，中堂建成楼房，后堂五层的主楼扩大与两横相接，构成四周高楼围合的更具防卫性的形式。

福裕楼的中堂与两侧的过水屋及前后厢房将内院分隔成六个天井，使空间层次更加丰富。内院中的中堂砖木楼阁精致华丽。

楼前是窄长的前院，院前的照壁紧临溪边，院门设在一侧，显然是风水上的原因，门楼旋转一个角度，斜对水口。其四周为二层至五层的土楼，夯土墙承重，墙面以白灰粉刷。整座建筑中轴对称，屋顶错落，气势轩昂。

The Fuyu Building, built in 1882, is located in Hongkeng Village, Hukeng Town, Yongding County, Fujian Province. It is a transitional type between the Wu Feng Lou and the Fang Lou (square-type building). The lower hall of the Wu Feng Lou develops into a two-story building here, connecting horizontal houses of three-storey on both sides; the central hall is built into a building and the five-storey main building in the back hall is expanded to connect with horizontal houses on both sides. Enclosed by towering buildings on all sides, the entire building acquires a more defensive form.

The central hall, the rooms on both sides, the front chamber and the back chamber together divide the inner yard into six patios in various sizes, enriching the spatial arrangement there. The central hall in the inner yard is made of brick and wood and constructed with pavilions, showing gorgeous beauty.

In front of the Fuyu building is a long and narrow front yard and the screen wall before it is close to a stream. Due to the feng-shui, its gate is set on one side and the door building is planned in a very special angle. Surrounding the gate are the earthen buildings of two to five stories, made of rammed earth walls which are painted with lime. Being of axial symmetry and with roofs arranged in an orderly manner, the entire residence presents an imposing momentum.

福裕楼鸟瞰 The bird's eye view of the Fuyu Building

福裕楼外观
The outward appearance of the Fuyu Building

福裕楼厅堂天井
The small yard in the hall of the Fuyu Building

福裕楼大门 The gate of the Fuyu Building

福建大田琵琶堡

The Pipa Castle, Datian County, Fujian Province

琵琶堡位于福建省西部三明市大田县建设镇建国村，又名龟头堡，因建在大山间的一座形似寿龟的山岗上而得名。堡始建于明洪武年间，历代有重修，因族人祖先信奉儒释道，故将土堡建成琵琶形。该堡由堡前代表琵琶琴弦的小溪沟、高石墙、夯土的厚实堡墙、坚实的门洞、畅通无阻的跑马道、应急的逃生窗、三圣祠、观音阁、中后楼等组成，是一座典型的依山形、重风水、喜宗教的异形土堡，具有良好的防御功能。

The Pipa Castle is located in Jianguo Village, Jianshe Town, Datian County, Sanming City, Fujian Province. It is also known as the Guitou Castle because it is built in a tortoise-shaped mountain (a meng is an animal that eats little but produces many eggs, like tortoise). The Castle was built during the reign of Emperor Hongwu of the Ming Dynasty and has been renovated in different periods. Since the ancestors of the clan believed in Confucianism, Buddhism and Taoism, the castle was constructed into a pipa-shape (a pipa is a plucked string instrument with a fretted fingerboard), consisting of brooks in front representing strings, high stone walls, high thick rammed-walls, doors made of strong stones, obstacle-free horse road, emergency escape window, Three-Saint Temple, Goddess of Mercy Pavilion, buildings in the central area and at the back, etc. It is a typical castle with a special shape and strong defense function, built along the slope of the mountain, giving importance to feng-shui and religions.

琵琶堡外貌 The appearances of the Pipa Castle

琵琶堡远眺 The Pipa Castle seen from the distance far away

大门入口 The entrance of the gate

首层檐廊
The eave veranda on the first floor

二楼檐廊
The eave veranda on the second floor

福建大田芳联堡

The Fanglian Castle, Datian County, Fujian Province

芳联堡位于福建省大田县均溪镇许思坑村，它建在高耸的大山余脉坡底的田边，为族人张氏宗亲于清代建造，历时5年完成。

芳联堡前方后圆，象征天圆地方。进堡前，经小道、双重护堡壕、半月池，进入堡门。堡门在左侧，入内有空坪，到上下堂、天井、后楼及护厝。平面布局功能分明，中轴对称，规整严格。四角有双重外凸的碉式角楼，是闽西三明市土堡中的唯一实例。其构造为夯土筑墙，三合土抹面，构造简朴，经济实用，重点装饰精美。

The Fanglian Castle is located in Xusikeng Village, Yunxi Town, Datian County, Fujian Province. It was built beside the farming field at the end range of the big mountains by the Zhang clan in the Qing Dynasty and it took five years to complete it.

The Fanglian Castle is square in front and round at the back, embodying the round sky and square land. Before entering the castle gate there is a small trail, a double trench and a half-moon pool. The gate is on the left and inside it are an open land, an upper hall and a lower hall, a patio, a building at the back and the protection house. The layout features distinct functions, axial symmetry and strict regularity. At four corners, there are dual-convex turrets, the only living example of the earthen castle in Sanming City of west Fujian. Constructed of rammed earth walls covered with mixed earth, the building is simple, economical and practical, with exquisite decoration.

芳联堡外观 The outward appearance of the Fanglian Castle

芳联堡庭院 The courtyard of the Fanglian Castle

厅堂天井 The small yard of the hall

围楼建筑 The encircled houses

天井 The small yard

过水廊 The cross-water veranda

福建大田安良堡

The Anliang Castle, Datian County, Fujian Province

安良堡位于福建省大田县桃源镇东坂村，它建在坡度较大的大山余脉的坡地上，分为三个大台阶，两边各由14级台阶状土台逐级构筑，为当地熊氏祖先于清嘉庆年间所建。建造者依安良除暴、平安如意之意，取名安良堡。

该堡前座为一字形的两层下堂房屋加横屋，正中为合院式建筑。进堡前先通过瘦窄的小道、深溪沟、独木桥、高台，再经双重堡门而入。堡内为大院落，经两侧台阶式木构建筑可登上堡顶上堂。堡内前低后高，两侧夯土筑墙包房，整体建筑朴实而雄伟，具有天然的防御功能。

The Anliang Castle is located in Dongban Village, Taoyuan Town, Datian County, Fujian Province. Built on the steep slope of the end range of the high mountains, it is divided into three big steps, with 14-step earthen platforms on both sides. It was built by the local Xiong clan ancestors during the reign of Emperor Jiaqing in the Qing Dynsaty. With a hope for fighting against the violence and being safe and peaceful, its founders named it Anliang (peace and smoothness) Castle.

The front of the castle is composed of a stroke-shaped two-storey house and horizontal houses. Its center is of a courtyard building. Before entering the castle, one has to pass by a narrow trail, a deep brook, a single-plank narrow bridge, the high walls and the double gates. Inside the castle is a large courtyard and the wooden-step buildings on both sides can lead to the top hall of the Castle. Lower in the front and higher at the back, the Castle is constructed with rammed earthen walls on both sides to surround the rooms. The whole architecture is simple and magnificent, providing a natural defense.

安良堡外观 The outward appearance of the Anliang Castle

土堡中轴厅堂鸟瞰 The bird's eye view of the axis hall in the earthen castle

远眺土堡 The earthen castle seen from the distance far away

后部围楼 The encircled houses on the back

厅堂 The hall

围楼室内 The interior of the encircled houses

土堡护厝 The side building of the earthen castle

土堡护厝 The side building of the earthen castle

福建永安安贞堡
The Anzhen Castle, Yong'an City, Fujian Province

安贞堡位于福建永安市槐南乡洋头村的山谷之中，坐西朝东，背靠大山，规模宏大。平面前方后圆，中轴对称。堡前有横向广场，两侧有护厝与入口，周边矮墙环护，广场前有半月形水池。

堡宅是一组两层两进院落式建筑，庭院两侧和各房之间由通廊联系。整宅外围是一圈前方后圆的两层护楼。护楼内侧有一圈层层抬高的通廊，护廊外侧是厚达4米的石砌夯土墙，上层一圈跑马廊联系着全宅的防卫系统，前方左右各设一炮楼。炮楼凸出墙外3米，有瞭望孔和射击孔，具有良好的稳定性和防御性。安贞堡的自然环境前低后高，并由一组石级层层向内引申，步步抬高，楼宅的屋顶与廊檐高低错落，组合巧妙。

The Anzhen Castle is located in the valley of Yangtou Village, Huai'an Town, Yong'an City, Fujian Province, facing the east and lying against huge mountains, being of a large scale. Square in the front and round at the back, it is of axial symmetry. In front of the castle is a broadwise square and on both sides are protection houses and the entrance. The Castle is surrounded by a low wall and there is a half-moon pool in front of the square.

The Castle is a two-storey two-courtyard building. The two sides of the courtyard are connected to each room through corridors. The castle is surrounded by a two-storey protective building which is square in the front and round at the back. Inside the protective building there is a ring of gradually elevated corridor and the outer side of the protective building is a stone rammed earthen wall with a thickness of four meters. Because it is decisive to the defense system of the entire residence, the Zouma corridor on the upper side is set with a gun turret at each end, which protrudes three meters out of the wall, equipped with watching holes and shooting holes, making it stable and defensive. Enjoying a natural environment of being lower in the front and higher at the back, the Anzhen Castle has a set of stone stairs leading inwards and upwards step by step. The roof and corridor of the Castle are placed in different elevations with a clever combination.

外观 The outward appearance

外墙 The exposed wall

主楼天井 The small yard of the main building

堂楼 The main building

护厝天井 The small yard of the side building

厅堂天井 The small coutyard of the hall

屋顶造型 The looks of the roof

屋檐直角处理 The right-angle treatment of the eaves

屋檐转角处理 The corner treatment of the eaves

堂楼 The main building

层叠错落的屋顶 The roofs distributed with distinct gradation

楼居 The buildings

大门 The gate

二楼厅堂 The hall on the second floor

天井 The small yard

后围 The enclosed house on the back

福建尤溪茂荆堡

The Maojing Castle, Youxi County, Fujian Province

茂荆堡位于福建尤溪台溪乡盖竹村，又称潹宅厝，建于清光绪八年（1882年）。土堡依山而建，居高临下，前为开阔的水田，景色优美。

建筑风格独特，融合土堡、围龙屋、围屋及当地民居等多个元素。平面布局合理，结构复杂，通道纵横交错。土堡坐东北朝西南，平面呈长方形，通面阔105米，通进深60米，占地面积约6300平方米。

建筑中轴线上从前至后依次建有下堂、天井、厢房、上堂等，两边各三幢护厝。堡内木雕、石雕工艺原始，雕刻精美，内容丰富，寓意深刻。

茂荆堡坚固的堡墙，保护着土堡四周，前方后圆、高大厚重。堡墙通高8.5米，厚3米，墙基部分有6米高，均用坚硬的大块毛石垒砌而成。堡墙中后部设置碉式角楼。

The Maojing Castle, located in Gaizhu Village, Taixi Town, Youxi City, Fujian Province, is also called the Jizhaicuo. It was built in the eighth year during the reign of Emperor Guangxu in the Qing Dynasty (1882). The Castle is built along the hillside, in a towering post with open paddy fields in the front, enjoying beautiful landscape.

It has a unique architectural style blending the features of the earthen castle, the Weilongwu (the dragon-encircled house), the Weiwu (the encircled house) and the local vernaculars. It has a reasonable layout, complicated structure and crisscross paths. Facing southwest, the Castle is in the rectangular shape, with a breadth of 105 meters and depth of 60 meters, covering an area of about 6,300 square meters.

On the central axis, the buildings founded from front to back are: the lower hall, the patio, the wing-rooms, the upper hall, and three protection houses on both sides. Wooden and stone carvings with original craft are also found inside the castle, which have rich content and profound meanings.

The sound castle walls, which protect the Castle from outside intrusion, are square in the front and round at the back, magnificent and heavy. With a height of 8.5 meters and thickness of 3 meters, the walls are constructed on a 6-meter-high foundation which is made of hard chunks of rubble. The walls in the rear of the castle are equipped with towers-type turret.

茂荆堡全貌 The full view of the Maojing Castle

前围 The enclosed house in the front

前院 The front yard

后围 The encircled houses

建筑底层干阑做法
The structure of the first floor of the buildings with crisscross-wood poles

过水廊 The cross-water veranda

敞厅 The open hall

福建闽清岐庐
The Qilu, Minqing County, Fujian Province

岐庐位于福建闽清县坂东镇溪峰村，为清代江西省九江知府张鸣岐故居。

闽清岐庐古寨堡，1858年完工，1862～1874年又加固扩建。外围护墙宽75.4米，深59米，皆由巨石垒砌，上筑里外两重土墙，每重厚0.7米，高2米，外、中、里门硬木板厚10厘米，重达200余斤，结构坚固。

寨堡内雕梁画栋，尤以木雕、灰塑为盛。厅廊屋宇梁架、屏风上刻有《水浒传》、《三国演义》、《孙子兵法》等历史人物故事雕画以及花草鸟兽等，神态逼真，栩栩如生，呼之欲出。

The Qilu is located in Xifeng Village, Bandong Town, Minqing County, Fujian Province. It used to be the residence of Zhang Mingqi, magistrate of the Prefecture of Jiujiang in Jiangxi Province in the Qing Dynasty, who was sworn brothers with Lin Ze-xu and Shen Bao-zhen, the former was the Governor of Guangdong Province and Guangxi Province and the latter the Governor of Jiangxi Province and Jiangsu Province as well as the Nanyang Trade Minister.

The historic Qilu castle of the Qing Dynasty was firstly built during 1842-1853 and completed in 1858. During the period of 1862-1874, the Castle was reinforced and expanded, with a width of 75.4 meters for the external wall and a depth of 59 meters. Being made of giant stone, the Castle is constructed with double earthen walls with thickness of 0.7 meters and a height of 2 meters for each. Its outer door, middle door and inner door are all made of hardwood with a thickness of 10 cm, weighing 100 kilos, boasting a solid structure.

The inside of the Castle is richly ornamented, especially the wooden carvings and plaster sculpture. The beams of the halls and corridors as well as the screens are engraved with historical figures and stories from Outlaws of the Marsh, Romance of Three Kingdoms, Military Law of Sun Bing, and flowers, birds, animals and other pictures, which are vivid, and lifelike.

岐庐大门 The gate of the Qilu

厅堂 The hall

厅堂檐廊 The eave veranda of the hall

江西龙南关西新围

Xinwei Weilong Building in Guanxi Township, Longnan County, Jiangxi Province

新围位于江西龙南县关西乡下九村，建于清嘉庆、道光年间，因从老围分迁此地新建，故称“新围”。新围为三层围屋，平面近方形，中轴对称，四角设炮楼。内院中，三组三进的单层合院大宅连排建造，颇有气势。

围屋底层外墙由三合土夹卵石筑成，二层为厚土筑墙，三层外墙由青砖砌筑，内隔墙均为土坯。炮楼高四层，歇山屋顶，屋内上部不设固定楼梯，只在战时加强防卫时，才临时架木梯登楼。

围屋内院中以墙分隔成前、后两区。前区为附属建筑，中段有花园。后区为主体建筑，合院大宅三进三排，共14个天井。宅内梁柱构架工整华丽，雕龙描凤，彩画镏金。

围屋大门设在前区的右侧，前区的左侧还有一个后门。整个内院布局井然、空间丰富、层次分明。

Xinwei Weilong Building, located in Xiajiu Village of Guanxi Township, Longnan County, Jiangxi Province, was built during Emperor Daoguang's reign in the Qing Dynasty. It is called “the New Weilong” because it is moved from the old place and rebuilt in a new place. The New Weilong is a three-storey architecture, with a nearly tetragonal plane and symmetrical central axis, and a blockhouse is set up on each of the four corners. The three-group, three-row, single-storey yard dwelling houses are built in successive rows in the Weilong building, which is rather grand with imposing momentum.

The exposed wall on the base of the Weilong building is constructed with tabia sandwiched with cobbles. The wall of the second floor is made of mud and that of the third floor is made of sun-dried mud bricks inside and laid dark bricks outside. The blockhouses are four stories tall with xieshan-style roof (roof with four sloping surfaces and nine hips). There is no fixed staircase on the top of the building. Only in the war time or on occasions of defense, contemporary wood staircases would be set up.

Inside the building, a wall is set up to separate the space into two parts: the front part and the back part. The front part stands the subsidiary architecture with a garden in the middle. The main architecture is located in the back part. The whole building has three rows with three lines, and fourteen small courtyards altogether. Inside the residence, the structural elements of the beam columns are tidy and beautiful, carved with dragons, drawn with phoenixes or gilded color paintings.

The main entrance is located on the right side of the front part, and a rear door is arranged on the left side of the front part. The whole interior yard has a neat and tidy layout so that the space is plenty and the gradation is distinct.

关西新围外貌 The appearances of the Xinwei in Guanxi Township

围内外院 The outer courtyard in the Wei

厅堂檐廊 The eave veranda of the hall

宗祠大门 The door of the ancestral temple

巷道 The lanes

江西龙南燕翼围
The Yanyiwei, Longnan County, Jiangxi Province

燕翼围位于江西省龙南县杨村乡杨村圩，为四层长方形围屋，始建于清顺治六年（1649年），历时五年建成。

围屋中心是长方形的内院，院内一角有一口水井。围屋每层共33间房，只设一部楼梯。二、三层为卧房，内侧挑出木构回廊，每层回廊木地板上留两个洞口，以便临时搭活动木梯上下。第四层为储物层，不设内回廊，而是在外围设环周通道，以便防卫时救援。

围屋底层外墙为厚2米砖墙，外包2.5米高的条石勒脚。二层以上为砖与土坯混合承重墙，内隔墙为土坯墙。二层以上每个房间均设内大外小的火枪眼。整个围子只有一个出入口，设三道门：第一道门为围门，是厚实的木板门扇，外包铁皮。第二道门是从上面落下的原木板闸，闸门上方还设朝下的射击孔和注水孔。第三道门才是日常使用的便门。围子南北两个对角对称地凸出四个炮楼，俯视围子，外形似展翅飞燕，故名燕翼围。

The Yanyiwei, located in Yangcun Village, Yangcun Town, Longnan County, Jiangxi Province, is a four-storey enclosed building in the rectangular shape. It was built in the sixth year during the reign of Emperor Shunzhi in the Qing Dynasty (1649) and completed five years later.

Its center is an inner courtyard in the rectangular shape with a well in one corner. There are 33 rooms on each floor which are accessible with only one staircase. Living rooms are arranged on the second floor and the third floor and the inner sides on these two floors are designed as wooden corridors. Each corridor has two holes on the wooden floor to lock the movable wooden ladder for temporary use. The fourth floor is used for storage without any inner corridor. However, its outer side is equipped with a circumferential path for rescue in case of the need of defense.

The outer wall at the ground floor of the building is the 2-meter thick brick wall, strapped with a 2.5-meter-high stone flag. Load-bearing walls above the second floor are made of mixed brick and earth and its inner walls are made of earthen walls. Each room on the second floor and above floors is built with perforation on the walls for shooting which is larger inside and smaller outside. There is only one entrance in the building with three doors: The first door is an enclosed one made of solid wood covered with metal. The second door is a board gate made of raw wood which unfolds from the above. It is equipped with shooting holes and water-injection holes on the upper side. The third door is for daily use. At the two corners in the north and south, there are four gun towers overlooking the entire place symmetrically. Since it looks like a swallow on full wing, the building is named Yanyiwei (swallow wing enclosure).

燕翼围内院鸟瞰 The bird's eye view of the interior yard of the Yanyiwei

围屋跑马廊 The Paoma corridor of the encircled houses

顶层围屋房间 The rooms on the top of the encircled houses

顶层围屋通道
The passage way on the top of the encircled houses

广东客家角楼

The Hakka Jiao Lou (turret), Guangdong Province

广东客家民居平面形式多样，组合灵活，其基本类型有门楼、锁头屋、堂屋、横屋等，也有一种叫角楼的形式。前者为客家民居中的基本类型，由堂屋、横屋组合而成的复合类型就形成了大型和中型客家民居。在大、中型客家民居周围加上围屋就形成了封闭型建筑，当地称为楼。楼一般为方形，通常称为方楼，也称为方围，也就是角楼的形式。它的特点是防御性强。有的地区称之为四角楼。

角楼的特点是将方楼或方围的四周房屋做成多层式，一般为三层到四层，在四角还做微凸的碉房，对外开设枪眼窗，作瞭望和射击用。楼的内部有回廊相通，外围设壕沟，通风采光靠天井。

The Hakka vernacular houses in Guangdong vary in flat forms and have flexible combinations. The basic types are: the gate building, the U-shaped house, the main room, the horizontal house, as well as a form called Jiao Lou (turret). A Jiao Lou is a basic type of the Hakka vernacular in large-and-medium size consisting of a main room and a horizontal house. In the medium and large Hakka house, the vernacular is reinforced with a weiwu to form a closed building, locally known as Lou (building). A Lou is generally in square shape, often referred to as Fang Lou (square building) or Fang Wei (square enclosed building), which is of the turret form. Characterized by strong defense, it is also known as Si Jiao Lou (four-corner building) in some areas.

A Jiao Lou features a square building or the surrounding buildings are of multi-storey house, usually three to four stories. At four corners there set convex towers with shooting windows opened outwards for outlook and shooting. Its inner buildings are connected with verandas with ditches set in peripheral side. The patio is for ventilation and lighting.

广东和平林寨镇兴井村角楼

The Jiao Lou in Zhenxingjing Village in Heping Stockade of Guangdong Province

兴井村中宪第内院

The interior yard of the Zhongxian Mansion in Xingjing Village

广东连平客家角楼瞭望窗
The observation window of Hakka Jiao Lou in Lianping County, Guangdong Province

广东连平油溪镇
黄氏大夫第内巷
The interior lanes of the Official Residence of the Huang Clan in Youxi Town, Lianping County, Guangdong Province

广东梅县
南华又庐
Nanhua Youlu Weilong Building, Meixian County, Guangdong Province

南华又庐位于广东省梅县南口镇侨乡村，是一座优秀的客家传统民居，由印度尼西亚华侨于清光绪三十年（1904年）建造。它占地一万多平方米，共有房间118间，大小厅堂几十个，当地称为“十厅九井”。

南华又庐坐西朝东，平面形式是：中间为堂屋，两边为横屋，横屋旁边是杂间和花园。宅居之前有禾坪、水塘，后面不是围龙屋，而是一排整齐的枕式屋。比较特殊的是两侧的横屋都是单元式布局，有厅有房，为各自独立的完整的庭院小屋，俗称“屋中屋”。整座围屋，中轴对称，布局规整，外观稳健，有一定气势。

南华又庐室内装修讲究，其木雕构件、屏风门窗的图案花饰丰富多彩。庭院内种植树木、花草，还布置有鱼池、凉亭等，是一座环境秀丽的民居建筑。

Nanhua Youlu Weilong Building, situated in Nankou Town of Meixian County, Guangdong Province, is an excellent Hakka's traditional vernacular architecture, built in 1904 (the thirtieth year in Emperor Guangxu's reign in the Qing Dynasty) by the overseas Chinese from Indonesia. With an area of more than 10,000 square meters, the building has 118 rooms and several scores of halls, big or small, which is called “ten halls with nine courtyards” by the local people.

Nanhua Youlu faces east. The main hall is in the middle and the sideways houses are on both sides, beside which are the lumber room and the garden. In front of the building are the sunning ground and water ponds. The back of the building is a row of tidy houses shaped like a pillow, rather than Weilong. Something special is that all the sideways rooms are of the unit layout, as independent houses, with halls and bedrooms of their own, which are called “houses in a house” by the local people.

The interior decoration of Nanhua Youlu is tasteful and exquisite, with rich and colorful patterns and ornamental designs on the wood-carving structural elements, the screen boards, doors and windows. In the yards grow trees, flowers, grass and fish ponds and kiosks are set up. Therefore, the building is really a beautiful vernacular architecture.

南华又庐远眺 The Nanhua Youlu seen from the distance far away

南华又庐门厅 The door hall of the Nanhua Youlu

广东大埔
张弼士故居
The Former Residence of Zhang Bishi in Dapu County, Guangdong Province

张氏光禄第位于大埔县西河镇车轮坪村，是著名华侨实业家、张裕葡萄酒"金奖白兰地"创办者张弼士的故居。张弼士于1903年授光禄大夫。

光禄第建成于1908年，坐北向南，前低后高，是一座三堂、四横、一围的围屋。屋前为禾坪，禾坪前有围墙和转斗门，禾坪东、西两侧建厨房和杂物间。外横屋与后围屋连接，与一般的枕头屋有所不同，横屋瓦面由前至后分五级层层错落，后围两头抹角略带弧形，形成长方形的围屋。外横屋和后围屋高二层，立砖柱，建内回廊，二楼设木栅栏阳台，硬山顶，瓦面出檐。

光禄第共有18个厅，13个天井，99个房间，前檐和堂屋的梁架、封檐板以及屏风等木构件金漆木雕考究，不失为艺术精品。

The Guanglu Mansion of the Zhang clan, located in Chelunping Village, Xihe Town, Dapu County, is the former residence of Zhang Bishi, a famous overseas Chinese businessman and the founder of the "Gold Brandy" Changyu Wine. When he was young, Zhang Bishi was engaged in business in the Southeast Asia. In 1893, he was appointed Consul General of Penang Island by Emperor Guangxu of the Qing dynasty and Consul General of Singapore in 1894. In 1903, he was promoted to a first-rank official and then granted as the Guanglu Minister.

The Guanglu Mansion was completed in 1908, covering a floor area of 4,180 square meters, facing south and lower in the front and higher at the back. It is an enclosed building with three halls, four horizontal houses and one enclosed house. In front of the residence is a sunning ground with enclosed walls and an exit in its front. On the east and west sides of the sunning ground are the kitchen and the utility room. Different from other pillow houses, the outer horizontal house is connected with the enclosed house in the rear and the horizontal house is divided into five layers from front to back which are arranged in an orderly manner. The two corners of the building are slightly curved to form a rectangular enclosed house. Both the horizontal house and the enclosed house at the back are of two stories with brick columns and inward corridors. The second floor has a wooden-fenced balcony, the roof being solid, tile surfaced and protruding at eaves.

In total, the Guanglu Mansion houses 18 halls, 13 patios and 99 rooms. Its front eave, beam in the main room, fascia boards, golden lacquer screens and other wood components are delicate and exquisite in both craft and art, making it an art masterpiece.

宅居侧院 The side yard at the center of the residence

入口门楼 The gate-building of the entrance

宅居厅堂 The hall of the residence

中厅檐廊 The eave veranda of the central hall

天井内院 The interior yard of the small yard

横屋天井 The small yard of the sidelong rooms

中厅天井 The small yard of the central hall

游牧民居

Nomadic vernacular houses

我国北方有广袤的草原，蒙古、哈萨克等民族的先民经过长期的狩猎生活，逐水草而居，创造了具有特色的民居建筑，称为毡包。

毡包，古称“穹庐”，又叫“毡帐”、“帐幕”，是一种毡木结构结合的活动式房屋。它构造简单合理，轻便耐用，便于拆装，也便于迁移，所用材料全部可以就地取材。它是一种最古老的装配式建筑，其骨架由统一参数的撑杆，蒙古族称“哈那”（一种木制可伸缩折叠的圆形网架墙），木制圆形天窗，蒙古族称“陶脑”，连接天窗与墙的椽条，蒙古族称“乌尼”等标准构件所组成。

毡包平面呈圆形，其上部绑扎成圆锥形，用皮条鬃绳绑牢，根据气温的高低在上面覆盖1～2层毛毡，用绳索束紧。在下部周围的圆柱形网架外，则围以一圈活动毛毡，掀开后可四面通风。地面铲去草皮，略加平整，铺牛粪煨燃，驱潮气，再铺砂一层，然后再铺上特制厚毛毡、地毯2～3层。

毡包包顶为圆形天窗，用于采光通风，并设置有绳索的毡子，白日开启，晚上封闭。包内生活、居住、待客和煮食都在一起，大门东南向，便于晨起纳阳和避北方寒风，中央为炉灶。

Northern China boasts vast grassland. The hunting life of the Mongolian ancestors living by water and grass over a long period has created a unique vernacular architecture, the Mongolian Yurt.

Mongolian yurt, called "yurt" in ancient time, also known as "felt canopy", "tent" or "felt bag", is a movable house made of felt and wood. With reasonable and simple structure, being durable and handy, it is easy for disassembly, installation, and relocation with all material available on the site. As the oldest assembly-style architecture, it is composed of standard components with uniform parameters including "Ha Na" (round stretch and folding wooden grid wall), "Tao Nao" (round wooden skylight), "Wu Ni" (rafter connecting the skylight and the wall), etc.

Mongolian yurt has a round plane. Its upper side is comprised of "Ha Na", "Tao Nao" and "Wu Ni", which are bundled by leather strip and mane rope to form a cone-shaped roof. According to the temperature, the roof will be sheltered by one or two layers of blankets tightened by ropes. The cylindrical grid at the lower part is encircled with movable blankets which can be raised up in summer for ventilation on all sides. The ground is spaded and leveled, covered with burning cow dung to expel humidity, paved with sand, then covered with special thick blankets and two or three layers of carpets.

The roof of Mongolian yurt is a round skylight, known as "Tao Nao". Used for ventilation and lighting, it is mounted with blanket which has cord and is opened in the day and closed in the evening. Routine activities, habitation, receiving visitors and cooking are all held in the yurt which has a door facing southeast to meet the sunshine in the morning and prevent the cold wind from the north. In its northwest corner, there sits a Buddha shrine and in its center places the stove.

蒙古包

The Mongolian Yurt, Inner Mougolia Autonomous Region

蒙古包内排气孔 The air-vernt in the Monglian yurt

蒙古包 The Monglian yurt

蒙古包多建于草原上地势较高地带。包内入口正面和左面是牧民日常活动和接待客人的地方，正面为主位。西北角供佛龛，龛前禁忌坐人。入口内右侧是妇女活动、居住和放置炊具的区域。墙边放置箱柜。入口左侧堆置燃料。中央为炉灶。

蒙古包是民族文化的象征，它的装饰有浓厚的民族特色。顶毡和包内的地毡用深色鬃毛缝制出云纹、回纹等几何图形，"乌尼"及"陶脑"涂红色，"哈那"内衬有蓝色布幔，"陶脑"为头顶上一轮太阳，"乌尼"为太阳四射的光芒。蒙古包的外表，雪白的毛毡围盖在圆锥骨架上，顶部罩有蓝、黄、红等颜色的布料做成的如意花纹，好似朵朵彩云镶嵌在美丽的绿色大草原上，呈现出一派大自然的风光。

As a symbol of Mongolian folk culture, the decoration of Mongolian yurt has strong ethnic characteristics. The carpets on the roof and ground are sewed with cloud and geometric patterns in dark mane; "Wu Ni" and "Tao Nao" are painted in red; the inner liner of "Ha Na" is blue curtain; "Tao Nao" looks like the sun over the head and "Wu Ni" the sunlight raying in all directions. Looking from the outside, with the snow white blankets covering the tapered grid framework and the roof sheltered by fabric with ru-yi patterns in blue, yellow and red, the yurt looks like colorful cloud dotting on the picturesque green prairie, unfolding breath-taking natural scenery.

草原蒙古包 The Monglian yurt on the grasslands

草原牧场 The pastare of the grasslands

草原蒙古包 The Monglian yurt on the grasslands

蒙古包骨架 The frame work the Monglian yurt

敖包 The Obo

藏族帐房
The Tibetan's Tent Houses

藏族帐房 The Tibetan's tent houses

居住在我国西藏高原的藏族牧民，长期畜牧在地势高爽，水草丰盛的牧区，他们生活和居住的场所，是一种可拆可装的帐篷式民居，当地称为“帐房”。

帐房是一种活动式房屋，可拆可装。帐内有一根或二根帐杆支顶，高约3米左右，形成攒尖式或起脊帐顶，四周各竖立杆，然后四角以毛索拉扯帐篷，形成四角或多角帐幕，帐脚篷布固定在地上。帐房设一门，门上悬护幕，帐顶顺脊处开一长形天窗，采光排烟，夜晚用护幕遮盖，防雨防风和防寒保暖。室内地面铺毛毡或兽皮，中心有石砌火灶。

The Tibetan herdsmen living in the Tibetan Plateau of our country have long been raise their livestock in the pastoral area with high topography and sumptuous water and grass. And the houses they live in are a kind of tent-type dwellings that can be taken apart and reassembled, called tent houses by the local people.

Moveable are tent houses that can be taken apart and reassembled. Inside the house there is one or two around 3-meter tall tent poles to support the top of the tent to make a point top or ridged top. Some upright poles stand on the four sides of the tent and then some rough ropes are used to pull the tent to make a quadrangle or multi-angle praetorium, with the tarpaulin at the foot of the tent fixed on the ground. A tent has one door on which a parapet curtain is hung. A long skylight is put in the ridge of the tent for collecting daylight and discharging smoke in the day. At night the skylight is covered with the parapet curtain for wing-proof, rain-proof and cold-proof. The ground inside the tent is spread with felt or hide, with a stone stove placed in the center.

藏北草原游牧帐房
The tent houses of the nomadic people in the savanna of the Northern Tibet

藏族帐房 The Tibetan's tent houses

新疆哈萨克族毡房
The Kazak's Yurts, Xinjiang Uygur Autonomous Region

哈萨克毡房　The kazak's yurts

哈萨克族主要分布在新疆伊犁哈萨克自治州境内，其民族主要从事畜牧业，兼营农业，活动之地大多在河谷山间的广阔草原上。当地有上好的天然牧场和肥沃农田，水草丰腴，林木清秀，瓜果飘香，粮粟丰盛。

哈萨克族牧民除了冬天居住在块石或土木建造的简易房屋内，大部分时间是在草原的毡房里度过的。毡房非常简便、轻巧、实用和牢固，它无论从选材、制作，还是从构件的设计和造型来看，都达到了精美的程度。

哈萨克族毡房建筑的艺术主要表现在栅栏墙架撑杆、顶圈、草帘、毡毯、围带和木门及其构件所组成的帐幕建筑外形艺术上。此外，以3～5家为一单元的哈萨克族牧民，同移共驻，生产生活在大自然草原中和清澈的小溪附近，宁静安逸，环境优美。

The Kazak people are mainly distributed in Yili Kazak Autonomous Prefecture of Xinjiang. They are mainly engaged in livestock husbandry and do farming as part-time business on the vast river-valley or intermountain pastureland. There are best-quality natural grazing land and fertile fields with plenty of water and grass, aromatic melons and fruits, green and beautiful trees, as well as sufficient grain crops.

The Kazak herdsmen spend most of their time on the pastureland except for in winter when they live in the roughly-built houses made of riprap rocks or mud and wood. Being light, practical, steady and easy to transfer, the Kazak's yurts have already reached the exquisite degree whether in material selection, making or in design and modeling of the structural elements.

The architectural art of the Kazak's yurts is mainly reflected in the external form of the stay-poles of the palisade wall frames, the top enclosure, the straw mats, the felts, the parapet curtains made of enclosed belts, wood doors and their auxiliary parts. In addition, three to five families of the Kazak herdsmen generally form a little community. They move together and settle down together, live and conduct their production on the natural pastureland near or by a clean and clear stream, with a quiet, comfortable and beautiful environment.

毡房入口　The entrance of the yurts

哈萨克毡房和毡房入口　The kazak's yurts and their entrance

哈萨克族毡房室内　The interior of the Kazak's Yurts

天山脚下的哈萨克族毡房
The Kazak's Yurts at the foot of the Tianshan Mountains

新疆塔吉克族毡房
The Tajik's Yurts, Xinjiang Uygur Autonomous Region

"塔吉克"是民族的自称，历史悠久，主要生活在新疆西南部帕米尔高原的塔什库尔干地区，以其勤劳、勇敢、淳朴、豪爽的性格和多姿多彩的民风习俗而著称于世。

塔吉克族的畜牧业占主要地位，过着半游牧半定居生活，春播以后上山放牧，秋季回村收获过冬。牧民在村中有固定住宅，一般为土木结构平顶屋。屋内不分间，四周筑有土台，为坐卧起居之地。在夏秋放牧季节才使用毡房，毡房中央有一灶，用以烧火、做饭和取暖。到了秋季转场时，毡房收起，就地放好，来年春天再搭起来居住。

The Tajik is what the ethnic group of people call themselves. With a long history, the Tajik people mainly live in the Taxkorgan area of the Parmirs in the southwest Xinjiang. They are famous for their diligent, brave, honest, bold and forthright disposition as well as their colorful folk customs and conventions.

The Tajik live a semi-nomadic, semi-settled life since animal husbandry takes a major place in their daily life. After the spring sowing, they put their cattle and sheep out to pasture on the mountains and go back to their villages for harvesting the crops in the autumn and pass the winter at home. The herdsmen have fixed homesteads, generally flattop houses with post and panel structure in the village. The Tajik people's houses are not divided into rooms. On the four sides within a house, a platform is constructed for people to sit or sleep on. The yurt is only used during the grazing period in summer and autumn, in the center of which there is a cooking stove for people to make fire, cook their meals, and warm themselves. When the time comes for them to move back home, the yurt is packed up and placed on the spot so that the owner can put it up and live in it again the following spring.

位于南疆的塔吉克族毡房
The Tajik's Yurts in South Xinjiang Autonomous Region

塔吉克族毡房 The Tajik's Yurts

新疆图瓦人牧房

The Tuwa People's Vernacular Houses, Xinjiang Uygur Autonomous Region

在新疆阿尔泰山深处的喀纳斯湖区，生活着大约2000多名图瓦人，他们世代以放牧、狩猎为生，居深山密林，沿袭传统的生活方式。

浓密的山林脚下是图瓦人居住的木屋，木屋采用井干式结构，坡屋顶。当地居民将砍伐下来的条木晾干、去皮、出榫、引槽，以井干式工艺直接搭成自己的居屋。

每家图瓦人的房前都有栅栏，同时用栅栏围出几个不同功能的院落，除了用于围住牛羊，也有用于生活、用于围住菜地等功能。

In the Kanas Lake area of the depth of the Altai Mountains, Xinjiang live the Tuwa ethnic group of people, about 2000 in number, who make a living by hunting and herding generation after generation. Living in the deep mountains and thick forests, they carry on as before the traditional life style.

At the foot of the thick mountain forests are the Tuwa people's blockhouses, with slope roofing and jinggangshi(no upright column and summer beam) structure. The local people dry by airing the barwoods, remove their peel, make tenons and mortises out of them and then construct their houses with the barwoods with the jingganshi techniques.

Every Tuwa family has a fence in front of their house and encircles the yard into several parts with the fence to serve for different functions. Apart from using the yard as a cattle-shed or a sheep-yard, the Tuwa people may use it as a living yard or for growing vegetables.

图瓦人木屋 The Tuwa People's wood houese

图瓦人毡房 The Tuwa People's Yurts

戈壁民居

Vernacular houses of the Gobi desert

我国新疆维吾尔自治区境内，阿尔泰山山脉与天山山脉形成的准噶尔盆地、天山山脉南麓的塔里木盆地，是我国西北地区最大的两个沙漠地带。维吾尔族人民长期居住在此，他们分布在新疆各地，绝大部分在天山以南的喀什、阿克苏、和田、库尔勒、吐鲁番和北疆的伊犁地区。他们的居住地有众多的绿洲分布，水源充沛，土质优良，农业作物和瓜果丰盛。

维吾尔族民居为适应沙漠地带的气候特点，就地取材，采用生土和生土成品建造厚实的外墙，以挡北方风沙。屋内用木构架和密梁满铺椽子的平顶结构，上铺厚草泥。平面围合式，内院周围为外廊，木柱拱形结构，廊柱蓝绿色。院内绿化，常栽植葡萄、石榴、无花果、桑等果树。这种民居，当地称之为“阿以旺”。

民居卧室多有坑床，周围墙面做成壁龛，用于摆放餐具物品。内墙壁面常用石膏雕饰，如印模压花、石膏雕花或木雕、彩画等。地板铺有毛毡，墙壁也挂悬毛毡，室内装饰华丽，色彩鲜艳。

阿以旺民居平面布局灵活合理，不强求中轴对称。内院的外廊是维吾尔族居民的重要室外活动场所。外廊木柱分为柱头、柱身、柱裙三段，分别装饰，两柱头之间，可单拱，也可双拱并连建造，券角部分做透空花饰，柱裙段与栏杆连接。所有木柱、栏杆、拱券都刷有不同颜色的油漆，如天蓝、红色、绿色，十分艳丽。

维吾尔族民居具有自由、幽静、简朴、亲切、舒适的风格，其建筑造型和装饰的特点是大量使用尖拱式或蒜头形，这是中国和中亚伊斯兰教文化交流的结果。

Within the Xinjiang Uygur Autonomous Region, the Junggar Basin formed by Altay Mountains and Tianshan Mountains and the Tarim Basin at the south foot of the Tianshan Mountains, are the two largest desert areas in northwestern China, where Uygur people have been settling for centuries. They inhabit in various parts of Xinjiang, mostly in Kashgar, Akesu, Khotan, Korla, Turpan at the south of Tianshan Mountains, and Yi Li area in northern Xinjiang. There are many oases in the places they live, boasting inexhaustible water, fertile soil, rich agricultural crops and fruits.

To adapt to the desert climate, the vernacular house of Uygur people is constructed with local material to block the sand wind from the north, using crude earth and finished earth products to build thick external walls. Inside, the house is built into flat-topped structure with wood frame and dense beam rafters paved with thick grass earth. The courtyard is of enclosed style with verandah around it, arched wooden structure and blue-green pillars. Ayiwang, the name given to this kind of dwelling by the local, is often planted with grapes, pomegranate, fig, mulberry and other fruit trees for greenery.

Inside the bedroom, there is usually pit bed with its surrounding walls made with niches for the display of tableware. The interior walls are often decorated with plaster carvings, such as embossed flowers, carved plaster flowers or wood carvings, paintings and so on. The indoor decoration for the house is ornate and bright with blankets covering the floor and hanging on the walls.

With a very flexible and reasonable layout, the vernacular house has no compulsory requirement for the symmetry of the axis. The verandah of the courtyard serves as an important location for outdoor activities of Uygur residents. Its wooden pillar is divided into three sections, chapiter, column body and wainscot, which are decorated separately. Between two chapiters, there can be a single arch or dual-arch and the corner place is decorated with hollow floriation; the wainscot is connected with railings. All wooden pillars, railings, and arch are painted in diversified dazzling colors, such as sky blue, red and green.

Presenting a style of being free, peaceful, simple, gracious and comfortable, the vernacular house of Uygur people is characterized by large use of arching or garlic shape in its architectural form and decoration, which is the result of the cultural exchanges between China and Islam in the Central Asia.

新疆喀什维吾尔族民居

The Vernecular Houses of the Uygur People, Kashi, Xinjiang Uygur Autonomous Region

喀什气候干燥、风沙大，居民喜近水建宅，因而构成了喀什街巷的自然曲折、密集幽深，柳树成荫、流水潺潺，民居高低错落的布局景象，有的在巷道上空搭建过街楼，更增加了喀什民居的迷人特色。

封闭式的喀什民居具有良好的保温性能。夏季烈日当空，民居庭院在树荫和外廊的遮蔽下，热空气不易侵入。傍晚，在微风中，民居内的热空气流动上升，新鲜的空气进入庭院和室内，民居内格外凉爽。喀什民居均设外廊，外廊在夏季可乘凉、就餐、待客，冬季可晒太阳。

民居的装饰重点是庭院的外廊和客厅室内。外廊木柱分为柱头、柱身、柱裙三段进行装饰。室内装修中，客厅墙面设许多壁龛，石膏制成，作摆放餐具茶具用。客厅顶棚下做有石膏花带或绘制彩画。

民居庭院中种植葡萄、石榴、无花果等，小庭院种植盆栽花木。夏日里枝繁叶茂、浓荫铺地，形成清爽的小气候。

It is dry, windy and dusty in Kashi and people here are fond of building their houses close to water, so the streets and lanes in Kashi are naturally winding, thick and deep, with willows shading the streets and running water murmuring. The layout of the vernacular houses with different heights, some buildings built over the streets, add more charming features of Kashi.

The enclosed houses in Kashi have very good warm-keeping qualities. In summer when the hot sun is high in the sky, it is not easy for the hot air to invade the vernacular courtyards since they are under the shade of trees and the lanai. But in the evening, the hot current in the houses rise up in the breeze and fresh air comes into the courtyards and houses thus become extraordinarily cool. All the houses in Kashi have lanais, under which people can enjoy the cool, have meals, and entertain their guests in summer, and bask themselves in the sun in winter.

The focal points for decoration in a house are the lanai in the courtyard and the interior of the drawing room. Three sections of a wood column (the head, the main part and the skirt-board) in the lanai are decorated respectively. In the decoration and furnishing of the interior of the drawing room, many niches are set up in the surface of the walls with plaster, used to put dinnerware and tea set. Plaster laces or color paintings are made under the ceiling of the drawing room.

In the courtyard, grape vines, pomegranate trees and fig trees are grown. In some small courtyards, potted plants are grown. In summer, the courtyard is shaded with luxuriant foliage and thick branches, and a cool microclimate comes into being.

喀什民居过街楼 The buildings over the street in Kashi

喀什民居聚落　The vernacular houses in Kashi

喀什民居街巷　The lanes of the vernacular houses in Kashi

喀什楼居　The buildings in Kashi

喀什民居檐廊 The eave veranda of the vernacular houses in Kashi

喀什民居室内壁龛装饰
The inner wall basket decorations of the vernacular houses in Kash

喀什民居室内石膏花墙饰
The plaster laces wall decorations in the interior of the vernacular houses in Kashi

新疆伊犁维吾尔族民居
The Uygur People's Vernacular Houses, Yili, Xinjiang Uygur Autonomous Region

伊犁地处新疆伊犁河上游山间河谷盆地，气温昼夜相差大，这里土地肥沃，水草丰美，风景优美。

伊犁建筑有强烈的地方色彩，其基本特点是大量利用生土、木材等土产材料，采用厚墙、高台、平坡顶、厚顶盖（草泥）、小窗、厚门等形式以适应当地自然气候条件和满足生活需要。

伊宁市塔西库鲁乡果园街的宅居最有特色。民居临街而建，大门后为院落，民居平面常建成"L"形或一字形，外有开敞的明廊，并由明廊连接每一个居室。明廊和明廊前葡萄架下的空地上，除刮风下雨和严冬飞雪的季节外，这里是居住者白天活动的主要地方。庭院形式变化多样，春到秋，绿叶拥簇、花果不断，充满着美好的生活气息。到了冬日，白雪盖地，居民们的生活移向室内，地毯艳丽，四壁挂毯、彩画、箱龛、被褥铺陈，暖意融融。

Yili is situated in the Hecong Basin on the upper reach of the Yili River, Xinjiang. The temperature here varies greatly between day and night but the soil is fertile with plenty of water and lush grass, and scenic beauty.

The architecture in Yili has a strong local color. Its basic character is to make use of plenty of such material as the immature soil, wood etc. thick walls, daises, flat roofing, thick top caps (mud with grass), small windows and thick doors are adopted to meet the local natural climate conditions and the need of living.

The houses of the Guoyuan Street in the Taxi Kulu Township, Yining City have the most distinguishing features. The architecture is built on the frontage in a row or in the shape of "L", and the courtyard is just behind the main gate. Outside the houses, there is a spatial corridor with which all the rooms are connected. The corridor and the open space under the grape shed is the place where the dwellers have their daily activities in the daytime except when a strong wind is blowing, when it is raining or when the snow is flying in cold winter. The appearance of the courtyard varies greatly. From spring to autumn, the green leaves form a cluster with incessant flowers and fruit, full of magnificent flavor of life. When winter comes, the ground is covered with white snow, and people move indoors for their daily activities, where the carpets are gorgeous, and the rugs on the walls, the colored paintings, the boxes and baskets and the bedding and mattress, fill the air with warmth.

民居外观 The outward appearance of the vernacular houses

马车 The carriage

伊宁楼居 The buildings in Yining

民居大门 The gate of the vernacular houses

檐廊 The eave veranda

民居室内陈设 The furnish of the interior of the vernacular houses

夏季常在檐廊置厨灶和休息垫床
The kitchen stove and mattress in the eave veranda in summer

民居院落 The vernacular courtyards

民居檐廊 The eave veranda of the vernacular houses

新疆吐鲁番维吾尔族民居

The Uygur People's Vernacular Houses, Tulufan, Xinjiang Uygur Autonomous Region

吐鲁番位于新疆东部火焰山脚下，夏季气温高于45℃，降雨量极少。寒冷气流使吐鲁番周围刮起八级大风，飞沙走石，是典型的沙漠干热型气候。长期以来，维吾尔族居民修筑坎儿井引水灌溉，并利用火焰山周围高强度微红色粘土创建了土拱、土墙的半地下两层生土居住建筑，抵挡了烈日高温的侵袭，形成了新疆特有的生土建筑文化。

吐鲁番民居一字形，楼上平面为一明两暗，明间小，两侧暗间大，其中较大者是主要卧室。民居单面采光，不讲究通风。半地下室由4个拱形空间组成，深30～80厘米，柱廊位于二层，是户外起居生活场所。

当地盛产葡萄，不少民居建有葡萄风干房。风干房全用土块镂空砌筑，形象半虚半实，是吐鲁番民居中特有的风貌。吐鲁番民居的庭院更是千变万化，庭园绿化中必种葡萄，并搭设棚架。

Tulufan is situated at the foot of the Flaming Mountain in the eastern part of Xinjiang, where the temperature in summer can be as high as 45 degrees Celsius and the rainfall is extremely rare. With the typical desert dry heat climate, the cold current causes force-8 wind in the area of Tulufan, carrying sand and driving stones. For a long time, the Uygur people there have been building the karez (an irrigation system of wells connected by underground channel) to draw water for irrigation. They also take the advantage of the high-intensity, reddish clay around the Flaming Mountain to build their two-storied, half-underground dwellings with cob arched top and cob walls to prevent the attack of the strong sunshine and the high temperature, presenting a unique immature soil architectural culture in Xinjiang.

The dwellings in Tulufan are arranged in a line. The upper floor of a house has three rooms, the middle room is smaller than the two side ones, and the biggest one is the bedroom. People living in such houses are not particular about ventilation since only one side of the house is well lighted. The semi-basement comprises 4 arched spaces, 30 to 80 centimeters deep. The columns stand on the upper floor, which is the place for people to enjoy their daily life.

Tulufan is rich in grapes and so many houses have grape-drying structures, which are built with hollowed-out soil blocks, combining void with solid as a peculiar scene of the vernacular houses in Tulufan. The courtyards there are infinite in variety but there are certainly some grape vines climbing the shed frame in a courtyard.

风干葡萄的凉房 The house for air drying grapes

院落葡萄架 The grape trellis of the courtyard

吐鲁番民居与二楼风干房 The vernacular houses in Tulufan and the grape-drying structure on the second floor

高原民居

The vernacular houses in the plateau mountainous regions

西部高原地区，由于受奇特多样的地形、地貌和高空气流的影响，形成复杂的独特气候，日照多、辐射强、气温低、温差大、干燥、多大风、气压低、氧气薄。此外，山区地势高低起伏多，垂直高差大，地势险要，交通十分困难，农田面积很小，一般分布在山腰或山麓一带。

在材料方面，高原山区盛产石料，建造房屋以石材为主，如毛石、片石、碎石以及被风化了的岩石——阿噶土（西藏特有的土质）都是本地最主要的建筑材料。木材中以柳木、杨木、松木为常用之材。

高原山地民居大多为内院式房屋，两层，四面为封闭厚实的围墙，对外基本不开窗，向内院则开窗。如多户居住，则内院四周房屋各成单独单元，有单独出入门口。每户各有木梯通向二层，互不连通。

民居墙体有用夯土者，但多数用石墙，墙基、墙裙也用石砌，因当地石材资源丰富，石砌技术又很高超，因而它就成为了高原建筑，也就是藏族建筑最主要的结构形式和藏族民居的特征。藏族民居通常建造在山麓或山腰台地上，厚实的墙体层层叠砌而上，故常用收分方法，外观下宽上窄，厚实稳重，给人一种屹立、挺拔、凝重、稳定的感觉，这就是高原地区垒石民居的特征。

色彩丰富是藏族民居特征之一。在建筑中常用白、红、黑三色，这是敬献给世界三层——天、地、地下之神的象征。如外墙毛石涂白色浆，外墙上檐有两圈红黑色带，门廊前有黑色矮墙等。

此外，门窗饰有上小下大的黑边框，大门及窗的上部有逐层出挑的彩画小椽和打褶“香布”小篷，都是寓意吉祥。粗犷的石质石材，本土的泥质白墙，飘柔的纺织品，各种材料的特点与色彩，和谐统一在藏族民居整体之中，它突出了藏族民居朴素、简洁而又豪放、稳健的性格美。

Due to a variety of unusual terrain, landscape and the impact of the high-altitude air current, the highland in western China has complicated and unique climate, featuring abundant sunshine, strong radiation, low temperature, big temperature difference, dry air, strong wind, low air pressure, and thin oxygen. In addition, because of steep mountains, uneven landform, large vertical height difference and dangerous terrain, transportation there is extremely difficult and only a few small plots of farmlands are found at the mountainside and the foot of the mountains.

As for material, the plateau area is rich in rock which is the main constructing material for house. For instance, gross rock, parcel stone, detritus, and weathered rock of A'ge soil (a unique soil in Tibet) are the most important building materials in the region. The timber of willow, poplar and pine trees are commonly-used building materials as well.

The vernacular houses in plateau area are mostly two-storey courtyard residence with enclosed solid walls and windows only opened inwards. Where several families live together, the buildings around the courtyard are constructed into independent units with their own entrances. Each unit has a wooden ladder leading to the second floor which is not connected with each other.

The walls of the vernacular house are built of rammed earth and mostly of stone, so are the wall base and wainscot. Thanks to the rich rock resources and eminent crafts and skills, the stone building becomes the principal structural form and the feature of the Tibetan architecture on the plateau. Usually built at the foot of the mountain or on the mountainside, Tibetan vernacular house is constructed with stones piling up layer upon layer to form thick and solid walls. Its appearance is characterized by wide foundation and narrow top, giving an impression of being towering, dignified and stable.

Rich color is one of the characteristics of the Tibetan vernacular house. The soil of three colors is commonly used in the construction, white, red and black, which embodies the dedication to the Gods of the Heaven, the Earth and the Underground. For example, the stone for external wall is roughly treated and painted in white, presenting a bold and unconstrained style. Along the edge of the external wall, two red-black ribbons of 5 cm encircle the entire residence. A black wall mounts before the door corridor.

Besides, the door and windows are decorated with a black rim which is larger at lower part and smaller at upper part. At the upper part of the main door and windows, there are small rafters decorated with colorful paintings, extruding layer on layer and the “fragrant-cloth” canopy folded into pleats with stripped textile. Both the small rafters and the “fragrant-cloth” canopy convey auspicious meanings. The rough stone material, white clay wall, silky textile, various materials and colors, all these are mixed in the Tibetan vernacular house in a harmonious and consistent manner, which give prominence to the plain and simple Tibetan house and reveal the robust and moderate characters of the Tibetan people.

西藏拉萨民居
The vernacular houses in Lhasa, Tibet

拉萨是一个由群山环抱的东西峡谷，海拔约3700米，是西藏政治、经济、文化中心，自治区的首府。

拉萨民居多集中在旧城区，以公元七世纪修建的大昭寺为中心逐渐向外扩展，形成八角街内外圈。八角街的内街小巷曲折狭窄，建筑间距小，密度大。建筑物彼此搭接，院落毗连，形成大院套小院，犬牙交错的总体布局。

总体布局很注重朝向，主要房间面向南、东两面，院落封闭，北西两面不开窗或仅开小窗以防风沙侵袭。民居多为外廊式二、三层楼房，以院落式居多。院落大小不一，少则几户，多则几十户。民居的结构为2米×2米正方形柱网组成的纵向框架承重体系，木梁柱，毛石或土坯外承重墙，粘土（垩嘎土）平屋顶。

Lhasa, an east-to-west valley surrounded by mountains with an altitude of 3700 meters, is the political, economic and cultural center as well as the capital city of Tibet.

The vernacular houses in Lhasa are mainly concentrated in the old city proper, spreading outward gradually with Jokhang Monastery as their center to form an octagonal street with an inner and an outer ring. The inner lanes within the octagonal street are both narrow and circuitous and the houses are built with little spacing and big density. The houses lap with joints and the courtyards are adjacent to each other to form a general layout of being interlocked just like dog's teeth with small courtyards situated in big ones.

Great emphasis is laid on the general layout of the vernacular houses in Lhasa with the main rooms facing either south or east. The courtyards are enclosed and there are no windows in the north and west walls or there is a very small window in it to prevent the attack of wind and sand. The houses are mainly two or three-story buildings with gabarits in courtyards. Some courtyards are big with several dozens of households in each of them while others are small with only several households in each of them. The structure of the houses is the longitudinal frame bearing system formed with 2 X 2 tetragonal prism webs, wood beam columns, ashler-made or adobe-made bearing walls and clay (Chalky GA soil) flat roofing.

街巷 The lanes

门檐 The door eaves

内院 The interior yard

拉萨八角街内巷民居
The interior—lanes vernacular houses of the Bajiao Street, Lhasa

外貌 The appearance

八角街民居窗饰
The window decorations of the vernacular houses of the Bajiao Street

内巷 The interior lanes

宅居外观 The outward appearance of the vernacular house

西藏波密米堆民居

The vernacular houses of Midui Village, Bomi County, Tibet Autonomous Region

林芝地区山高谷深，河流湍急，村寨大部分坐落在山脚、河边的小块平坝上。由于这里海拔相对较低，雨量较为充沛，适宜树木生长，它为林芝地区的民居建筑提供了木材。因此，林芝民居多为石木结构，也有单一的木结构。

林芝波密米堆民居位于米堆冰川脚下，建筑外墙大多为石砌，墙体材料除碎石、片石、卵石之外，木板、竹篱、柳条篱亦为多见。民居底层饲养牲畜或作储藏间，二层住人。屋顶为坡顶，将长条板子整齐地倾斜排列在楼顶上，用石块压住，作为建筑封顶。

In Linzhi Prefecture, the mountains are lofty and the valleys are deep and the current there is swift. Most of the villages or stockades are situated at the foot of a mountain or on small pieces of flatland by the river. Due to the relative low elevation, the rainfall is rather plentiful, thus the environment of Linzhi is suitable for trees to grow with particularly favorable natural conditions to provide the people there with sufficient wood for their house constructions. That's why the houses in this area are mainly of stone and wood structure, or simply of wood frame.

The vernacular houses of Midui Village, Bomi County, Tibet Autonomous Region are located at the foot of the Midui Glacier, and most of the exterior walls of the houses are built by laying stones. Apart from broken stones, flags, cobbles, wood planks are also used as the material for the construction of the walls, of course bamboo or willow twigs fences are often seen as well. Livestock are raised on the first floor that may also be used as storeroom and people live on the second floor. The roofing is of slope made of long battens arranged tidily in the tilt pressed by stone blocks as the binding.

米堆藏居 The vernacular houses of Midui Village

米堆藏居 The vernacular houses of Midui Village

米堆木结构藏居 The Tibetan vernacular houses of wood structure of Midui Village

米堆藏居外观 The outward appearance of the vernacular houses of Midui Village

米堆藏居 The vernacular houses of Midui Village

四川丹巴藏居

The Blockhouses of the Tibetan Nationality, Danba County, Sichuan Province

在素有“千碉之国”称誉的丹巴，碉楼主要集中在河谷的两岸。碉楼或三五成群，或独立山头。碉楼与碉楼之间依山就势，相互呼应，而集中的地方，一眼望去，几十座碉楼此起彼伏，连绵不绝，形成了蔚为壮观的碉楼群。

碉楼一般与民居寨房相连接，有的也单独筑立在平地山谷之中。碉楼的外形一般为高耸的方形柱体，从四角、五角到八角应有尽有，也有多达十三角的。

丹巴的山寨，以前称作碉楼和寨房。它们原本是两类不同性质的建筑，但随着时间的推移，碉楼和寨房已有机地结合为一体。碉楼寨房一般有三层，也有四层的，其一侧大多配有厢房，其顶层的外缘都围绕着黑、白、黄三种色带。

嘉绒藏寨中最具特色的是几百幢民居依山势而建，错落有致地融于自然环境中，体现了古代先民天人合一的传统理念。

Danba County has long been known for “the kingdom of a thousand blockhouses” that are mainly located on both sides of the Guhe River. Some blockhouses are grouped in threes or fives, some stand independently on the tops of the hills, but they echo each other since they are built according to the natural terrain with different elevations. Seen from a distance, several dozens of blockhouses, joining together, rise one after another in an unbroken line, presenting a splendid sight of a blockhouses complex.

The blockhouses are generally connected with the vernacular houses though some of them are built independently on a piece of flatland or in a valley. Generally with a rectangular column as the external form, as far as the column is concerned, the blockhouses have every form under the sun, from quadrangle, pentagon, to octagon, and some even have thirteen corners.

The fortified mountain villages used to be called blockhouses or stockade-house, which originally were two kinds of architectures with two completely different qualities. But as time goes by, the two have merged into an organic whole. The houses are generally of three stories, some of them four stories. Most houses have a wing room built on one side of it. The outer edge of the overstory, whether a blockhouse or a stockade-house, is enclosed with black, white and yellow belts.

The several hundreds of well-proportioned houses in accordance with the natural terrain have the most typical characteristics of the Jiarong (one of the Tibetan dialects) Zang people’s fortified mountain villages, symbolizing the ancient people’s traditional conception of harmony between man and nature.

梭坡藏居　The vernacular houses in Sopo Township

梭坡藏居群 The block houses of the Tibetan Natronality in Suopo Township

梭坡碉楼 The blockhouse in Suopo Township

梭坡碉楼群 The blockhouse complex in Suopo Township

四川丹巴梭坡藏居
The vernacular houses of the Tibetan Nationality in Danba Suopo, Sichuang Province

甲居藏居外观
The outward appearance of the vernacular houses of the Tibetan Nationality in Jiaju

甲居藏居大门
The gate of the vernacular houses of the Tibetan Nationality in Jiaju

甲居藏居室内
The interior of the vernacular houses of the Tibetan Nationality in Jiaju

高低错落的甲居藏族民居
The vernacular houses of the Tibetan Nationality distributed with distinct gradation

甲居藏居入口
The entrance of the vernacular houses of the Tibetan Nationality in Jiaju

嘉绒藏族服饰
The clothing of the Tibetan Nationality in Jiarong

甲居藏居独木梯
The "Single-Wood" ladder of the vernacular houses of the Tibetan Nationality in Jiaju

甲居藏居檐廊
The eave veranda of the vernacular houses of the Tibetan Nationality in Jiaju

四川丹巴甲居藏族民居
The vernacular houses of the Tibetan Nationality in Jiaj

甲居藏寨建筑群
The architectural complet of a Tibetan blocade in Jiaju

参考书目

[1] 侯幼彬.中国建筑艺术全集第20卷宅第建筑（一）北方汉族.北京：中国建筑工业出版社，1999，5.

[2] 陆元鼎.陆琦.中国建筑艺术全集第21卷宅第建筑（二）南方汉族.北京：中国建筑工业出版社，1999，5.

[3] 杨谷生.中国建筑艺术全集第22卷宅第建筑（三）北方少数民族.北京：中国建筑工业出版社，2003，10.

[4] 王翠兰.中国建筑艺术全集第23卷宅第建筑（四）南方少数民族.北京：中国建筑工业出版社，1999，5.

[5] 陆元鼎，杨谷生.中国民居建筑.广州：华南理工大学出版社，2003，11.

[6] 孙大章.中国民居研究.北京：中国建筑工业出版社，2004，8.

[7] 颜纪臣.山西传统民居.北京：中国建筑工业出版社，2006，3.

[8] 陆琦，唐孝祥，廖志.中国民族建筑概览（华南卷）.北京：中国电力工业出版社，2007，9.

[9] 戴志坚，李华珍，潘莹.中国民族建筑概览（华东卷）.北京：中国电力工业出版社，2008，5.

References

1. Hou Youbin. The structure of dwelling houses 1: The architecture of North China. Volume 20 of *The Complete Works of the Chinese Architectural Arts.* Beijing: China Architecture & Building Press, 1999.

2. Lu Yuanding, Lu Qi. The structure of dwelling houses 2: The Han people in South China. Volume 21 of *The Complete Works of the Chinese Architectural Arts.* Beijing: China Architecture & Building Press, 1999.

3. Yang Gusheng. The structure of dwelling houses 3: The minority nationalities of North China. Volume 22 of *The Complete Works of the Chinese Architectural Arts.* Beijing: China Architecture & Building Press, 2003.

4. Wang Cuilan. The structure of dwelling houses 4: The minority nationalities of South China. Volume 23 of *The Complete Works of the Chinese Architectural Arts.* Beijing: China Architecture & Building Press, 1999.

5. Lu Yuanding , Yang Gusheng. *The Vernacular Architecture of China.* Guangzhou: The Press of South China University of Technology, 2003.

6. Sun Dazhang. *A Study on the Vernacular Architecture of China.* Beijing: China Architecture & Building Press, 2004.

7. Yan Jichen. The Traditional Vernacular Houses of Shanxi Province. Beijing: China Architecture & Building Press, 2006.

8. Lu Qi, Tang Xiaoxiang, Liao Zhi. *A Sketch of the National Architecture of China:* the Volumc of South China. Beijing: Press of China Electricity Industry, 2007

9. Dai Zhijian, Li Huazhen, Pan Ying. *A Sketch of the National Architecture of China:* the Volume of East China. Beijing: Press of China Electricity Industry, 2008.

后记

本书著者为华南理工大学建筑学院教授。

翻译者为华南理工大学外语学院教授、讲师。

摄影者为陆琦，其中蒙古包照片资料由郑崴天提供。

浙江温州地区部分文字资料由温州市城乡规划设计研究所丁俊清高级规划师提供。

Postscript

The authors are professors of the Architectural College of South China University of Technology.

The translalors are a professor and a lecturer of the School of Foreign Languages of South China University of Technology.

The photographer is Lu Qi while the photos concerning.Mongolia are provided by Zheng Weitian.

The word material concerning Wenzhou, Zhejiang Province are provided by Ding Junqing, Senior designer in the Town-and-Country Planning Institute of Wenzhou City.